P9-CEM-280

You'd Better Believe It!

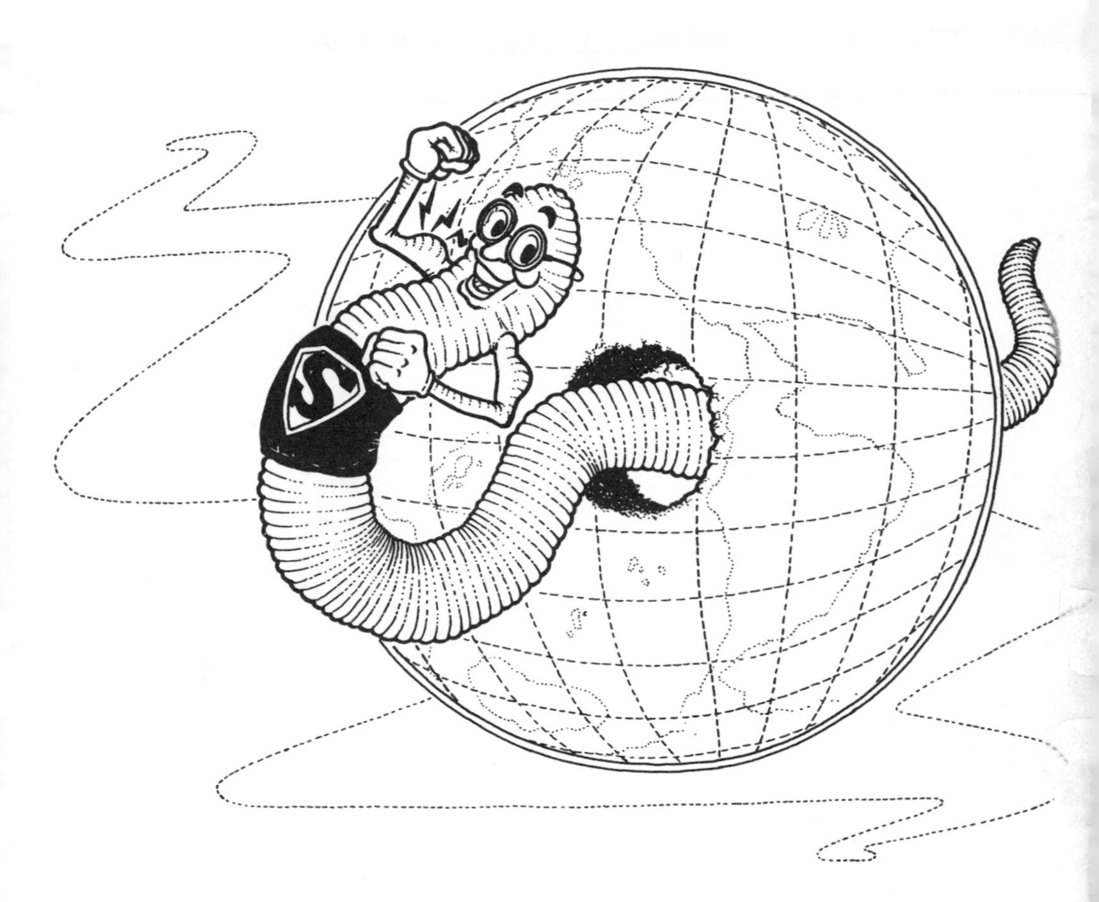

You'd Better Believe It!

by MARTIN M. GOLDWYN

Illustrations by Ted Enik

CITADEL PRESS Secaucus, N.J.

I dedicate this book to
STEVEN
LANI
and
STACI
with love

First edition

Published by Citadel Press
A division of Lyle Stuart Inc.
120 Enterprise Ave., Secaucus, N. J. 07094
In Canada: George J. McLeod Limited
Don Mills, Ontario
Manufactured in the United States of America

Library of Congress Cataloging in Publication Data

Goldwyn, Martin M
You'd better believe it!

SUMMARY: Answers such questions as "Can animals talk?", "How do we see?", "How long can bacteria live?", and many others.
1. Science--Miscellanea--Juvenile literature.
[1. Science--Miscellanea. 2. Questions and answers]
I. Enik, Ted. II. Title.
Q163.G63 ' 500 79-15160
ISBN 0-8065-0672-5

CONTENTS

PREFACE

Science, reflected in everything around us, comes from a Latin root that means knowledge. We are all born with inquisitive minds and can enjoy the fascination of scientific information if it is presented simply. But somehow science often makes people cower—enough to make students drop a course because they feel it is too technical. As a teacher, I have always been concerned with how scientific principles could be taught in a matter-of-fact, appealing manner. I do not mean that basic laws and principles should be compromised, but that they could be made more accessible to those who show an interest but who do not wish to delve into the more intricate aspects.

Having taught science in high school for some years, I have heard students ask hundreds of questions that ranged far afield from the work at hand, like "Why does sand come in different colors?" "Why does paint peel?" "Why can't ashes burn?" "Where do seashells get their colors?" "Why do mosquito bites itch and swell?" "Why don't birds get electrocuted?" and "Is there anything that is immortal?"

To determine which questions were most frequently asked, I began to collect and record the most thoughtful and serious ones that arose over the years. By making these part of the course, and setting aside time for informal discussion, I hoped to let the students see the relevancy of science to their lives. In time I was able to gather a variety of questions covering a broad range of topics.

Another source of questions for class discussion emanated from the hundreds of science fair projects I sponsored. For example, I learned from a girl who is now practicing medicine which animal made the strongest glue in the world. A student whose father had a multivaried fruit tree provided me with the question, "Can a tree grow more than one kind of fruit?"

It was gratifying to see this question-and-answer approach so well received by my classes, and from there it was only a logical progression to consider gathering the material into a book. I've attempted to maintain a style that is both light and accurate, adding a touch of humor wherever I could. In many instances I was required to capture the essence of difficult concepts while at the same time keeping the technical language to a minimum. However, a few topics remained which required the use of some technical terms in order to retain their accurate scientific meaning.

How important is science to our lives? Pick any object around you and consider how some phase of science entered into its development and manufacture. Though it may seem to us that the world is in constant flux, and, to paraphrase Thomas Wolfe, we can't go home again, in reality the laws of nature—be they physics, chemistry, botany, biology or whatever—remain unchanged, and it is the work of scientists to uncover their secrets. Our enlarged understanding has enabled us to create such twentieth-century phenomena as space exploration, lasers, and exotic chemicals and antibiotics. But nevertheless, the laws of nature are immutable, and it is this constancy that enables us to make scientific progress. For instance, we can rely on the fact that from primordial time to the present, two atoms of hydrogen and one atom of oxygen unite to form one molecule of water. The chameleon has been changing its color and fish have been surviving in the freezing Arctic for millions of years, while earthworms, roaches and many insects have likewise survived the millennia.

But it is we who are only now recognizing the plan of their existence and learning to understand it.

To find answers to these questions I consulted books in the fields of general science, biology, physics, astronomy, earth science, marine biology, medical and scientific journals and reports, and many other sources. Uppermost in my mind was a desire to reach anyone, young or old, with a sense of inquiry. I hope I have achieved my aim.

To the many friends in the various fields of science who have reviewed the contents and helped edit, I am very grateful. I owe special thanks to Burt Bond, a teacher of physics, biology, and marine science, whose encouragement and advice is most gratefully acknowledged. And thanks also goes to my wife Goldie for her continued patience and help throughout this entire project.

ARE THERE ANY ANIMALS THAT CANNOT REPRODUCE?

If it were possible for you to get confidential with a talking mule, he would probably tell you that he has no pride in his ancestors, and no hope for posterity. He is the offspring of a male donkey and a female horse, which is the usual arrangement in order to obtain in the offspring the best qualities of each parent. The horse is larger, a lot more handsome, and learns more easily. On the other hand, the donkey is more disease-resistant, sure-footed, and can perform work under conditions that would make a sensible horse rebel. However, and unfortunately, all male mules are born sterile, and with a few exceptions, the females are also sterile. But there is one dubious consolation for the mule. He does not have to worry about the mule population explosion.

HOW DOES A CAMEL GO WITHOUT WATER FOR DAYS?

The secret lies in the hump on his back. Contrary to common belief, it does not contain a crooked spine. It's a solid mass of energy-producing fat, weighing about 80 pounds or more. This is a food reserve station or storage house for the energy to be expended on long trips over hot desert sand. When the camel is starved, the hump almost disappears. The fat in the hump is used up while energy is created, but in this process of change, for every pound of fat about a pint of water is produced. Thus, by excreting very little, and by being one of the few animals capable of recycling its liquid body waste, it is able to go for days without water. Thus you see that an Arab used-camel buyer, about to make a long journey, would probably check the size of the camel's hump to insure good mileage capability, whereas you and I would check our cars for gas, water and oil.

HOW DOES A CHAMELEON CHANGE ITS COLOR?

If you were an insect that stumbled within the long-tongued reach of the chameleon, I would warn you not to fall for its quick color-change act. If you believed that the chameleon had left the arena, you would become an insect of the past tense. What the chameleon does to achieve a quick color change is to manipulate the pigment cells of its body called melanocytes. They can remain spherical to give him a light color, or he can expand and branch them out to give him a darker color. Octopus, cuttlefish, flounder and squid can do the same.

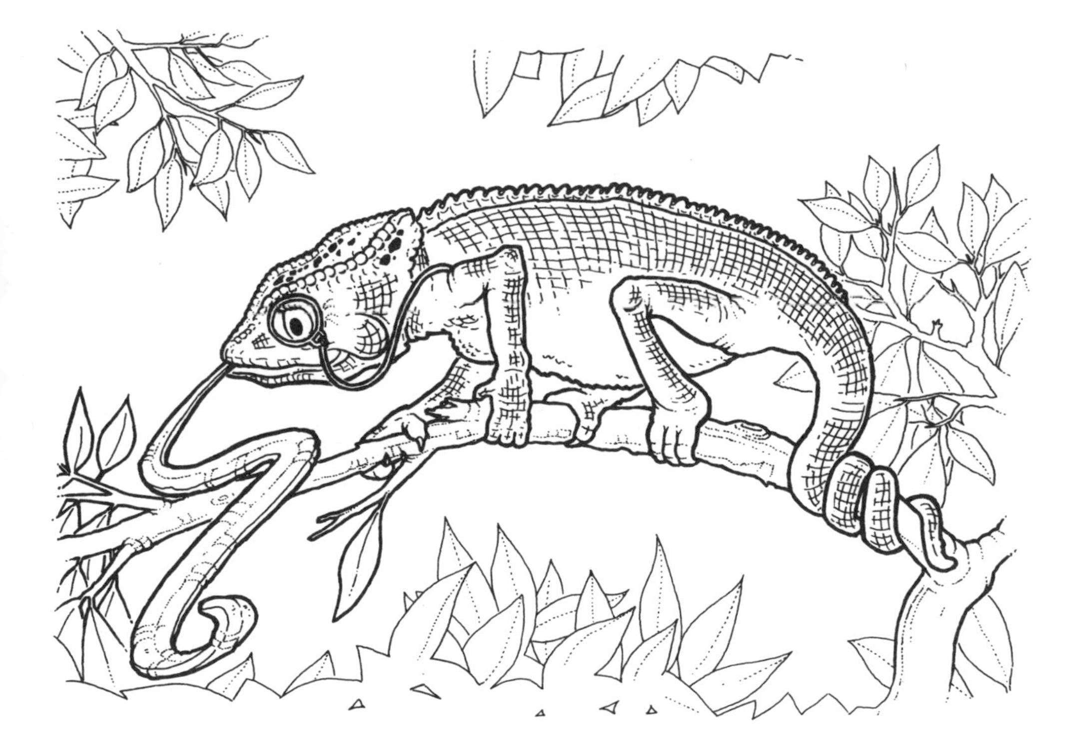

HOW DO OWLS SENSE THEIR PREY?

The ruff of feathers around the eyes of some owls is a sensitive sound receptor. This ruff covers the large ear openings of about 525 different owl species in the world. They all have a look-alike appearance with a large broad head and the display of feathers around the eyes.

Some of these predatory birds are day hunters, but most are night hunters that catch mice, rats, shrews and snakes. It should not amaze you that owls can swoop down upon their prey in the darkness of night and catch any night prowler rustling in the leaves and branches as long as their sound receptors are in good working order.

HOW DO FISH SURVIVE THE SUBFREEZING COLD OF ANTARCTIC SEAS?

If you think you have trouble starting your car in subzero weather, what should the fish of the Antarctic seas do with temperatures 5.4 F. degrees below freezing? These hungry fish need other fish to survive. But how can they catch them if they become stiff with cold? Mother Nature found the solution in glycoprotein, a substance discovered in these fish by scientists not long ago. This pure white substance has been isolated in polar fish, and has been compared to our radiator antifreeze. Any other fish without this substance placed in the polar sea freezes to death.

Whether this substance is made in response to a subfreezing condition as a natural body product is still unknown. What is definitely known is that nature only guarantees its proper function until the fish is swallowed by another.

WHAT IS THE OLDEST KNOWN SURGICAL OPERATION?

If you were an Aztec Indian, an Inca of Peru, or an African tribesman, you'd think twice before seeing your tribal doctor about a headache. He would grab you and start a trepanning operation to open your skull to let the demons out. When he was through with you, you'd really have a headache. Trepanning got its name from the instrument used, but surgeons no longer use it. Openings have been found in skulls of humans who lived thousands of years ago. It is therefore the oldest operation we know about. Certain tribes of North Africa and Melanesia still practice it when they think that one of their tribesmen is insane and they wish to release the demons.

However, with no background in science, no sanitary conditions nor antiseptic tools, very few of these primitives survived. You might say they needed this operation like a hole in the head.

WHAT ANIMAL MAKES THE STRONGEST GLUE?

Barnacles, which are related to lobsters, shrimp and crabs, make the strongest known glue. Out of about 800 species, most are classified as fouling organisms, because they attach themselves for life with their self-made cement to pilings, rocks, hulls and ships, which remain their permanent home. Paleontologists have traced their history back 400 million years and some fossils were found which had attached themselves 150 million years ago. Today, scientists are trying to analyze and duplicate the powerful cement they make.

Only a few countries use barnacles for food. Chile, Greece, Spain and Italy harvest and eat the goose barnacle. However, there is one barnacle found on the west coast of the United States, known as *Balanus nubilus,* which reaches a weight of three pounds and a height of five inches, and tastes much like lobster and crab.

HOW FAST DOES PAIN TRAVEL?

Our nervous system has over 10 billion nerve cells in a network that covers every square inch of our skin and organs. Like insulated electric wires the cells in the central nervous system have a similar protection to prevent the leakage of an impulse from its proper pathway. If you should slam your finger with a hammer, the sensation of pain will reach the brain before you can blink an eyelash. The speed of an impulse, including that of pain, varies considerably in humans. But it has been found to be able to travel as fast as 350 feet per second, almost one-eighth the speed of a bullet. Considering the millions of cells involved in this internal action to get the impulse to the brain, this is breathtaking. Millions of dendrites have to contact millions of axons to accomplish this. The answer from the brain will tell you to pull your finger away from the offending hammer, but the advice usually comes too late.

CAN ANIMALS TALK?

Among the animals there are mating songs and talk, food-finding signals, distress and danger shrieks, hissing and howls and expressions and calls of hunger. Animal behaviorists reported cats with a vocabulary of 17 "words" all of which are varieties of the familiar "meow." Two young male chimpanzees were taught a 36-word vocabulary to enable them to convey thought to each other in a sign-language way.

The sheep baas when hungry and the hen clucks for her chicks to come and join her in a find of food. The rooster lets everyone know he's the master of his surroundings. Tree animals let you know when you're invading their privacy. The leader of a pack of gray wolves searching for prey has three different signals to give them. There is a rallying call, a call that indicates a fresh scent, and a signal to close in. A crow discovering an owl gives out a strident series of sharp caws to alert and assemble her friends to join in mobbing the enemy. There are many instances to show that the animal kingdom knows there is strength in numbers.

When a herring gull discovers a rich source of food, it gives a special food-finding call that attracts other gulls. But when small quantities of food are found, the gull usually consumes them with no advertisement.

HOW DOES A FLY WALK UPSIDE DOWN?

Invisible to our eyes, but not to the compound eyes of the fly, the ceiling is a veritable expanse of Grand Canyon or Sahara Desert, full of valleys, ridges, hills and dales. To overcome these obstacles, nature has endowed the fly with all the proper equipment to grasp and hold this surface with each of its six legs. Each leg has a pair of tiny claws that resemble those of a lobster, and underneath the claws, a pair of small weblike fuzzy pads called pulvilli. These are functional suction pads which the fly presses to the surface to squeeze out the air and create enough suction to hold itself up. Thus, with its claws and suction pads the little pest can walk majestically upside down.

DO SOME PEOPLE HAVE MORE BONES THAN OTHERS?

Yes, many people do. At birth we have about 300 bones, and by the time we are adults there are only about 206. During the process of our growth and development, a great deal of bone fusion takes place. Five bones fuse to make the sacrum, and four bones fuse to make the coccyx. This partially accounts for the reduction in number.

However, not infrequently the process of fusion overlooks some bones, and therefore some persons may have an extra bone or two. We know, for instance, that five percent of the people have an extra rib in their rib cage. This happens more frequently in men than in women.

WHY DO VOICES OF MEN AND WOMEN DIFFER?

In the human voice box, known as the larynx and located in the throat, there are strings called vocal cords. These strings give sound to the voice when vibrated by wind from the lungs. The deeper voices of men are due to the longer and thicker cords in the voice box. That's what makes the difference between voices of men and women. Vocal cords keep growing until about age 13 in boys, when they become full grown. A boy's voice takes a sudden change when he reaches maturity because the greatest amount of growth takes place at that time. It is then that he feels the sudden breaking of the voice. Men over 65 may also have a voice change, taking on a high pitch and quaver.

OF WHAT USE IS GARBAGE TO MANKIND?

For quite a time some European countries have used garbage for fuel. But since coal and oil were formerly cheap in the United States, we never thought of using garbage. Now, with an energy crisis, science is trying to find a cheap way of turning garbage into energy.

Methane is an odorless burning gas and is the main ingredient of natural gas. Since it is formed in decaying organic matter such as garbage, we are making a serious study of its potential as a source of energy. Estimates say that each person contributes about a ton of garbage a year, which in terms of energy is equal to half a ton of coal. We shall try to put this to use. Today, sixteen plants are in operation, twelve are under construction, and many more are in different stages of planning.

Nashville is planning to use garbage to heat its government buildings, and St. Louis has plans for a citywide garbage system to use all the waste of its two-and-one-half million people. In Milwaukee, a pilot plant to make methane gas from solid waste is being studied. Chicago is already using steam from garbage incinerators to heat and cool government buildings. In Los Angeles, an experimental setup is producing enough electricity to supply 350 homes from the city's garbage.

HOW DO YOU EXPLAIN NONFLYING BIRDS?

It is elementary that birds are supposed to fly, but there are many birds that do not fly. They are the ostrich, penguin, kiwi, emu and the cassowary. All these descended from flying birds that lived thousands of years ago. Among the nonflyers that were hunted out and are now extinct are the dodo and the auk.

The flightless cormorant, a large, voracious sea bird, with webbed toes and a pouch to hold its catch, became flightless because of the ease with which it caught fish. Thus, through evolution the wings became shorter and shorter. Now it uses its wings solely as a means of balance.

WHAT MAKES BIRD SONGS DIFFER?

The sound of birds is produced by a voice box containing a flap of tissue called the syrinx. It is located at the lower end of the trachea just as it forks into the two

bronchi. Air passing this flap forcibly makes it vibrate. The shape and size of the flap help determine the quality of the song. Muscular control of the tension of this membrane allows the bird to control the pitch, which may vary from a simple call to a melodic song.

The hoarse caw of the crow is made by vibrating a short stubby flap, while the thrushes and warblers have a longer, more slender syrinx. Some birds are excellent mimics of sound, like the mocking bird who can faithfully mimic the songs of other birds. The parrots and mynas can mimic human speech, and the mynas, especially, can enunciate words with surprising delicacy of inflection.

One myna was so perfect in the mimicry of its owner's language that the owner had to get rid of him out of shame.

HOW DO WE REMEMBER?

How people can recall and remember is still a mystery to scientists, and we are only left with the results of a few experiments and the theory expressed by men in the field. We all know that our brain can not only store information gained in childhood, but can recall a large part of it in great detail.

One neurosurgeon in Montreal, Dr. Wilder Penfield, performed brain surgery where parts of the brain were naturally exposed. While the patient was fully conscious he stimulated different parts of the brain with two electrodes, and discovered that by touching a certain spot he could get the patient to recall in great detail certain things that occurred in the past. He concluded that there is hidden away in the brain a record of the stream of consciousness.

Some scientists think that certain chemical compounds such as ribonucleic acid (RNA) may contribute to the cells' retention of events. But how this stream of events is stored away subject to recall at a very distant future date is still the greatest medical mystery.

WHAT DO THE MOST BEAUTIFUL FROGS CONCEAL?

The colorful frog legion of Costa Rican fauna seems almost limitless. These midget marvels average about two inches in size and identified species already number more than one hundred. Both sexes are glamorized in the most spectacular colors and patterns and are rarely identical. Brilliant orange, yellow, red, green and black predominate in the most varied and gorgeous designs.

A differing garb results when the pigment cells, called chromatophores, expand and contract with changes in heat, humidity and light. These color cells, buried under the skin, remain unaltered during molting, and a frog with a new coat retains its elegance.

But just as the beautiful rose has its thorns, these beauties of nature have their toxins. Through their skin, underneath which are located hundreds of minute poison glands, they secrete a most toxic poison that acts on the nerves of mammals and birds and causes paralysis and quick death. Pre-Columbians and later aboriginals knew how to apply these poisons to a blowgun dart to bring down a monkey, a deer or a bird.

Today these poisons are valuable tools in modern neurophysiological research. Since the toxins affect nerve and muscle in certain ways, they can be used to analyze the various steps in the process by which nerve impulses are transmitted. Also, since they affect the heart, scientists hope to find a control for rapid, irregular heartbeat called fibrillation, or for increasing the strength of the pulse in failing hearts.

ARE THERE ANY USELESS STRUCTURES IN OUR BODIES?

Yes, there are a few. Of all the useful bones in our body—about 206—there's just one little bone that serves no useful function at all. It is a small bone known as the coccyx, located at the base of the spinal column. It starts out as four little bones that become fused into one by maturity. But it is just as useless as one bone as it was as four. Today its only function seems to be to support the theory that says that this bone is the vestige of what was once a tail when we walked on all fours tens of thousands of years ago. Pictures of an occasional child born with a tail are on record.

Another useless structure in our body is the troublesome appendix. This too is considered a vestige of a once useful part of the digestive tract.

There is also a small membrane in our eye called a nictitating membrane. This membrane functions in birds and mammals, but in humans it is reduced to an inactive fold at the inner corner of the eye. Some day, through evolution, it may entirely disappear.

The pineal body, an outgrowth of the roof of the brain, baffles scientists with the mystery of its activity.

CAN ONE TREE GROW MORE THAN ONE KIND OF FRUIT?

We all know that every seed has its future fruit, flower, or vegetable predetermined genetically. Nothing and no one in the world can make an apple seed produce an orange.

However, the question asks if a tree can grow more than one kind of fruit and the answer is a definite yes, but with two provisions. First, the branches of the different fruit trees must be properly spliced onto the original tree, and secondly, the different fruit of all the branches to be spliced must belong to the same genus.

A good example would be the experiments with the citrus fruits. Many orange trees in Florida have had branches of grapefruit, lemon and lime trees spliced carefully, and all four citrus fruits grow beautifully on the single tree. Remember, however, that all of these fruits belong to the citrus group.

HOW DO BIRDS FIND THEIR WAY DURING MIGRATION?

Homing pigeons, as well as some other birds, find their way by celestial navigation. Experiments in a German planetarium by Dr. Fritz Saur showed conclusively that starlings steer a course on their flyways by means of Polaris, the North Star. Dr. Saur made the

North Star in the planetarium appear due south and the birds followed it, even though they were flying in the opposite direction to which they had to migrate during that season of the year.

But what may apply to starlings may not apply to the other birds. The truth is that science is still not certain about all the aspects of birds in flight. While some say that birds fly by means of Polaris, others think that birds have the ability to sense magnetic fields. But whatever it is that guides them, they are smart enough to know that cloudy or rainy nights are no nights to be out on a flight. They stay grounded till the skies are clear.

HOW DID THOSE FOREST TREES GET PETRIFIED?

About 150 million years ago, a forest of trees fell into a swamp in which the water contained great amounts of dissolved quartz. This is a rock-forming mineral. The water soaked into the empty cells of the decaying wood. As time passed, the quartz hardened and took on the form and shape of the trees. The streaks and spots of yellow, red, purple and black were caused by the oxides of iron and manganese in the water. Thus the logs, which became solid stone, to this day show every detail of the original wood structure. Even the rings of growth that tell the age of the tree when it died can still be counted.

HOW DOES NATURE MAKE A CAVE?

There are some caves made by nature that we seldom get to see. Such is the ice cave formed in icebergs and glaciers by the melting and freezing ice. There is the earthquake cave, formed by water seeping into cracks, and the lava caves in the sides of volcanoes, formed when top lava layers harden, while lower layers still flow. A more familiar cave is the sea cave, formed by centuries of wave action against the sides of mountains. One of these, the famous Blue Grotto in the Bay of Naples, is visited by thousands each year.

The underground caves are formed in two ways, both by the action of water. In one, an underground stream finds its way through cracks in the rocks and causes enough erosion to form a cave. In the other case, water seeps down through a mountain containing limestone. As the limestone dissolves, the water carries enough of the mineral through the roof of the cave to form limestone icicles called stalactites and stalagmites. These limestone caves are famous worldwide attractions. Some even contain small lakes inhabited by blind fish adapted to the darkness.

In certain caves in France and Spain, the outlines of animals have been found painted and carved on the stone walls. In many caves that were inhabited by cavemen over 35,000 years ago, the bones of animals were found on the floor—a vestige of prehistory.

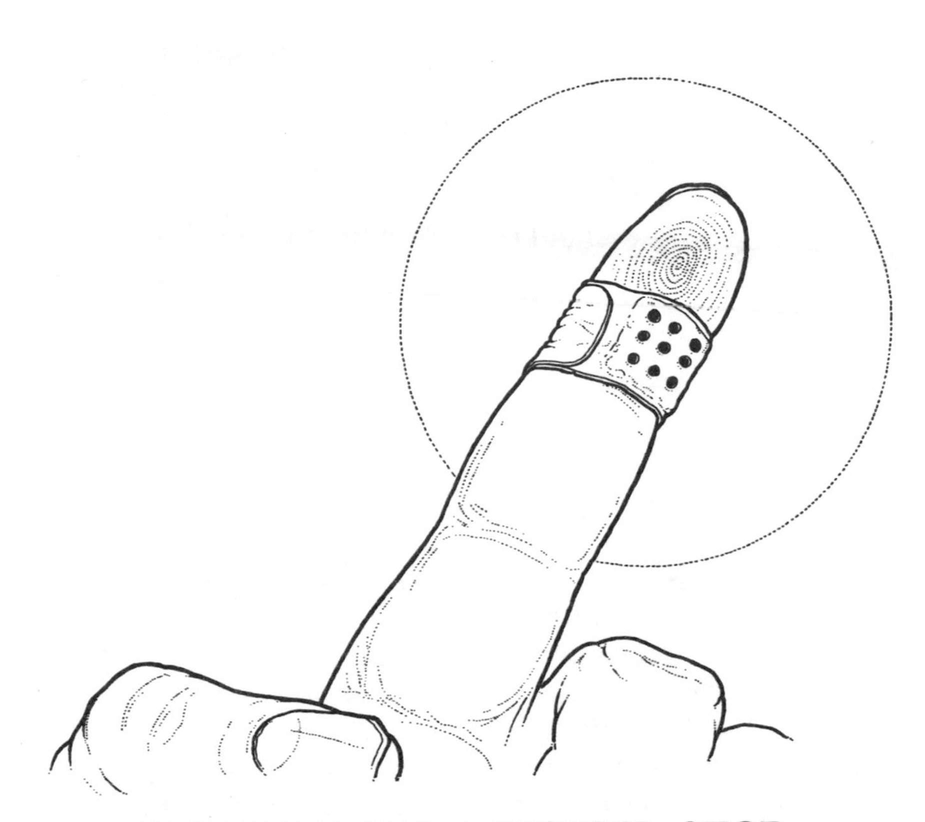

WHAT MAKES A WOUND STOP BLEEDING?

Our blood contains an enzyme called thrombin, which scientists believe is made in the liver. But the only time it becomes active is when there is an open wound. In the blood plasma there is a certain protein called fibrinogen. In an open wound the enzyme thrombin unites with the fibrinogen to form needlelike crystals called fibrin. This forms a network which catches blood corpuscles as they try to go through, thus creating a plug called a blood clot. This is nature's Band-Aid.

There are certain people who are born with an inherited trait which interferes with this process. Even a tiny wound can be fatal to the victim. This is known as hemophilia or bleeder's disease, which was common in many royal families of Europe. Although females carry the trait, only the males have the disease.

HOW DOES A PELICAN USE ITS POUCH?

A great big bird is the pelican.
His bill holds more than his belly can.

Not true! Not true! Let us tell it as it is. The pelican is an excellent diver who has been known to unite in a group to swoop down and drive fish into shallow water, where they can be caught more easily and in greater numbers. Then it uses its pouch to scoop up the catch, but not to hold or store it. The victims are immediately dropped into the stomach, where digestion takes place. Soon, the noisy baby pelicans, who by maturity will lose their voice, will be rewarded with mouthfuls of partially digested fish, freshly regurgitated.

DO TEARS ALWAYS MEAN SADNESS?

A happy dog exuberantly runs around in circles with his tail wagging a mile a minute. A cat will purr and purr and rub her back against you in extreme contentment, but not a hair on her face will disclose her happiness. But dogs, kittens and many other animals will whine and cry as any child would, though not a tear will show it. Only the voiceless dog, an African breed called Basenji, will have tears in its eyes when it cries.

Sea turtles, crocodiles and many marine birds also cry very salty tears that have nothing to do with emotions. Their tears are only the excess salt which they excrete by special tear glands in order to desalinate the salty ocean water they drink.

"Crocodile tears" is an expression meaning insincere sorrow. Old Croc isn't sorry. He's only thirsty.

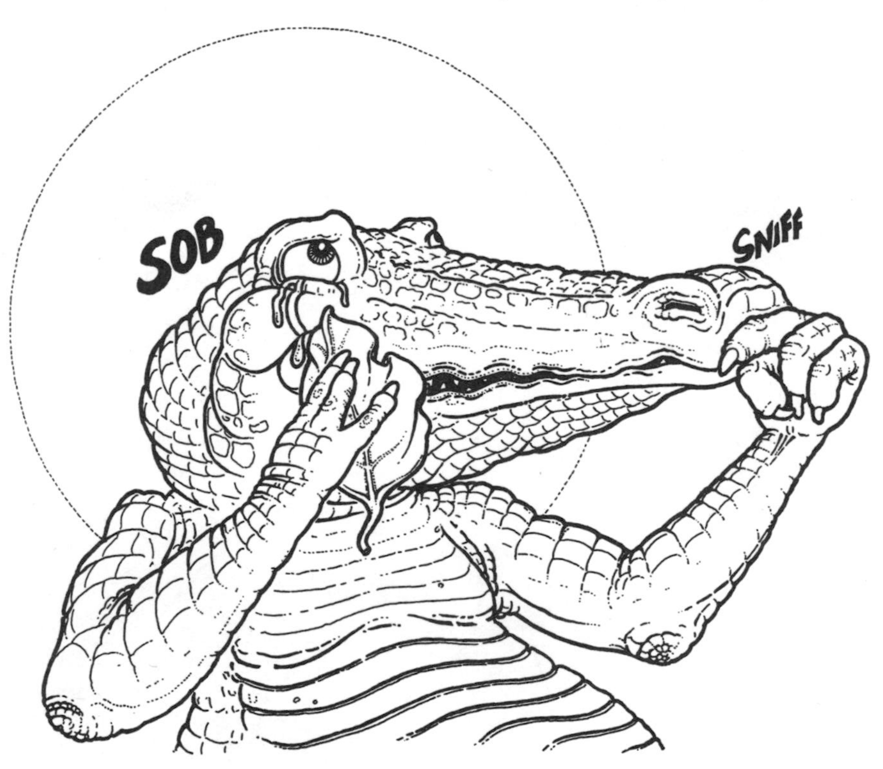

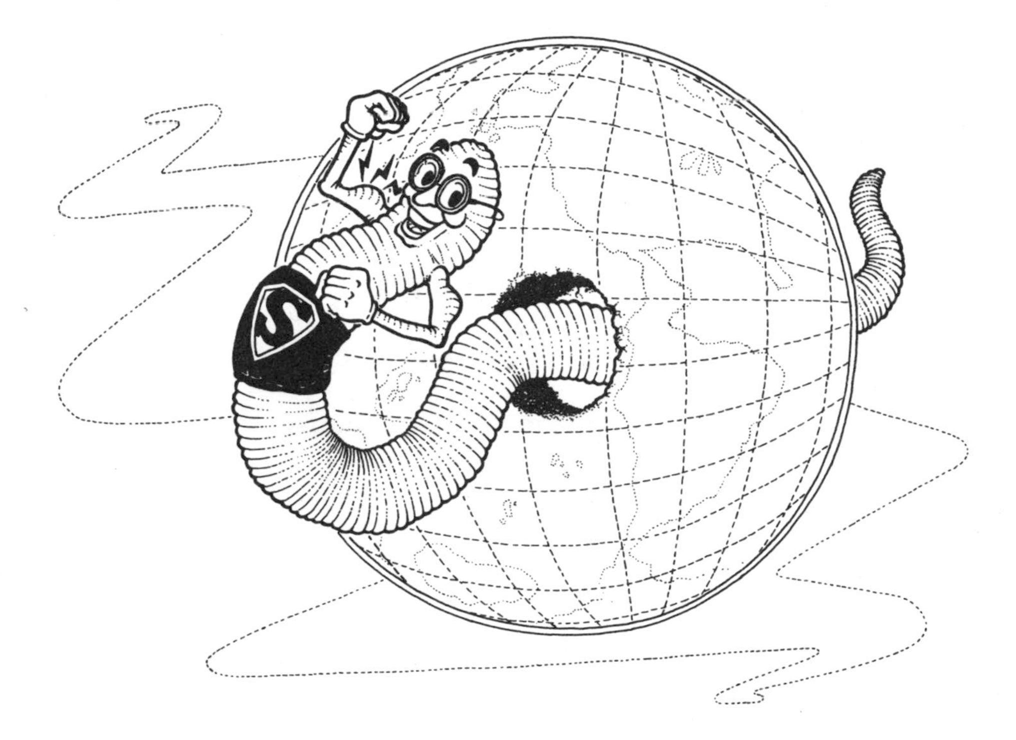

WHAT GOOD DO EARTHWORMS DO?

To get his food, the earthworm takes soil into his body and passes it out in a finely pulverized state called a cast, which is an excellent fertilizer. It has been estimated that earthworms produce about ten tons of cast to an acre of topsoil. These sightless worms range in size from one twenty-fifth of an inch to several feet long. They can cut their way through the hardest layers of earth and create burrows in the ground to give the soil better ventilation and moisture.

If you really wish to know the value of earthworms to our living world, just ask any fishing enthusiast.

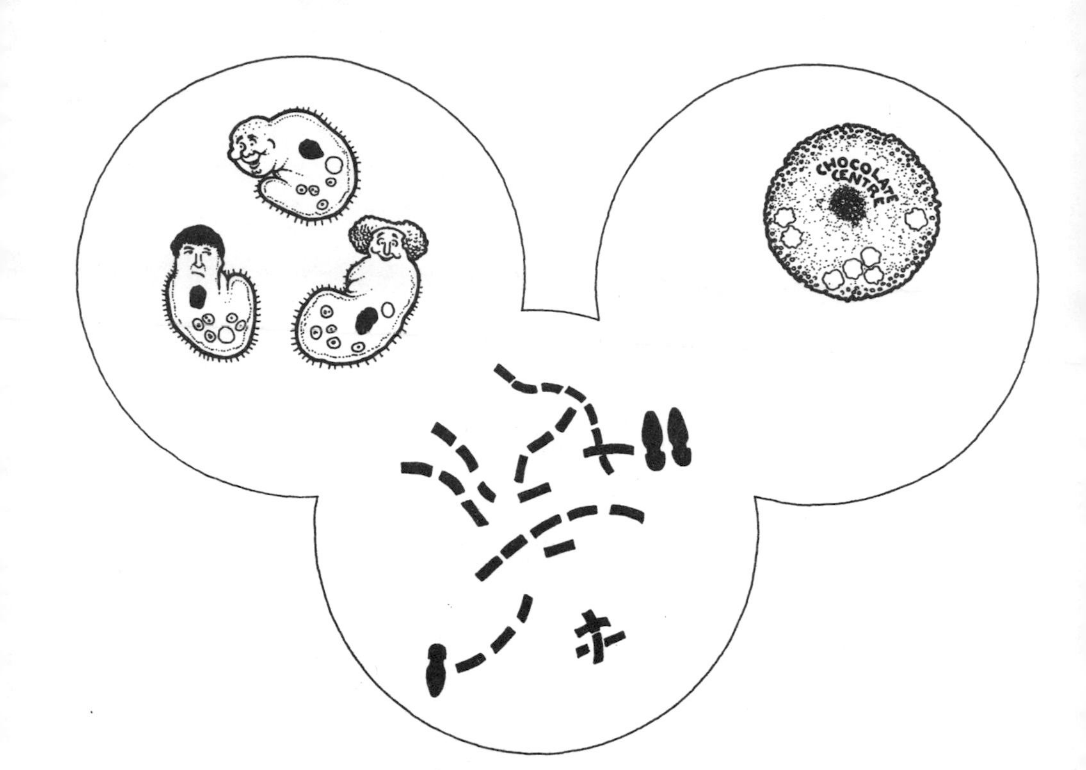

ARE THERE ANY BACTERIA THAT HELP MANKIND?

This world contains a variety of bacteria, some of which are flat, spiral, club-shaped, round, and some shapeless and constantly changing. Like the world we live in, some are goodies and some baddies. A fair guess would be that the good that the goodies do far outweighs the evil of the baddies. Here are a few examples of the goodies: Some turn the bodies of dead plants and animals back into the vital earth they came from. The nitrogen-fixing bacteria take nitrogen from the air and in their mysterious way turn the gas into solid nitrates and food we eat. These are found on the roots of peanut plants and the bean plants known as legumes. Bacteria in milk turn it sour to prepare it for a substance from which cheese is made. Leather, to be tanned, must be soaked in a solution containing bacteria.

In preparing linen and hemp, another solution containing bacteria is used. Many chemicals of industrial and commercial value are made by controlled bacterial action. Some good examples are formic acid, acetic acid (vinegar), lactic acid (in milk), butyric acid (butter), various sugars, ethyl and butyl alcohols, and acetones, which are fat solvents.

WHAT IS THE OLDEST LIVING THING ON EARTH?

The oldest living thing on earth is the world-famous General Sherman tree, a forest giant now growing in Sequoia National Park. Nearly 4,000 years old, it's as ancient as the Egyptian pyramids.

There are two giant species of this evergreen, the giant sequoia and the redwood. The redwood is the taller of the two, growing as high as 360 feet. This is over 36 stories high. Although General Sherman is over 27 stories high, it has a diameter of 36 feet at the base.

HOW WAS COAL FORMED?

Theory has it that coal began to be formed about 250 million years ago, when tree ferns about 60 to 80 feet tall fell into muddy swamps. Then the process of decay changed the cellulose into peat bogs which later became covered with inorganic material. The pressure of the overlying strata, together with the residual decay, increased the temperature in the peat bogs, slowly turning the peat into coal. Coal made in laboratory experiments seems to uphold this theory. Further proof are the imprints of the leaves of the giant tree ferns that appear now and then on the face of blocks of coal at the mine. The difference in the quality of coal is due to the variation in the amount of overlying strata and the subsequent rise of temperature.

WHAT MAKES THE DUCKBILLS DIFFERENT?

A duckbill is a four-legged, fur-bearing, warm-blooded mammal that gets its name from the shape of its nose, which is exactly like a duck's bill.

Why is it different? Well, all the other mammals give birth to live young, but this rebel (and also the spiny anteater) defy conformity in nature and lay eggs instead. It lays from one to three thin-shelled eggs about one inch long in a tunnel which it digs with its sharp-clawed toes. However, duckbills do nurse their young like other mammals. When the eggs are hatched, mama duckbill draws the babies to her with her paddle-shaped tail to nurse them. Several months later they all leave the tunnel.

Duckbills have webbed toes and are expert swimmers. They live along streams in Australia and Tasmania. If you are near them when a male is around, you had better watch out. He has a hollow claw or spur on each hind leg, connected to poison glands with which to scratch and poison his enemies.

You may know the duckbill by its other name, platypus.

HOW HAS NATURE HELPED BIRDS TO FLY?

A bird has fewer and lighter bones than other animals, and the framework is braced like the skeleton of a skyscraper. Many bones contain pockets of air connected directly to the lungs, and the wing muscles are attached to the keel of the breastbone for the best leverage. The feathers are insulators and conservers of heat, and are streamlined to permit smooth flight. The smaller ones fill the hollows of the body to keep friction down, and are arranged in overlapping rows like shingles on a roof. The main feathers of flight, the primary and secondaries, are attached directly to the bones of the wings.

WHAT'S INVISIBLE YET CAN PASS THROUGH CONCRETE WALLS?

Radio and TV beams can pass through concrete walls as X-rays pass through the body. And just as light rays bend when going through glass or water, so do radio and TV signals. This causes bad reception on indoor antennas and can be corrected by a tall outdoor antenna. Still the mystery remains how those skinny little radio and TV beams pass through concrete walls. Those high velocity particles called neutrinos have even greater penetrability and can go through the earth. Cosmic rays are also high penetrators and it was they that helped locate the secret crypts in certain Egyptian pyramids.

WHAT ARE THE SPAWNING HABITS OF THE SALMON?

There are many ways in which different male fish fertilize the eggs of the female. Each has a special way which nature has carried on for millions of years. The salmon is one of them.

Each year the adult salmon leave the ocean to fight their way upstream for several months to reach the lakes that form the headwaters of the stream. Then the female scrapes her tailfins to make a five-foot nest on the bottom and deposits about 10,000 pea-sized eggs. Immediately the male sheds his milt or sperm over them for fertilization. In a few days after this adventure, they both die.

The latest study shows that salmon follow the waters for spawning by their sense of smell. Out of 50,000 tagged salmon, 92% arrived in the river they originally came from. The few that strayed were found where salmon had been before. None were found where salmon had never been. This shows that salmon are attracted by the smells of other salmon and not just by the fresh water.

The salmon you eat did not spawn and die. Those spawning salmon bacame too sick to be used for food and died in a few days. Your canned salmon was a healthy one caught in the sea. Sometimes nature is hard to understand. Many animals enjoy their young, care for them, play with them, and fight to death for them. Here the salmon go through all that trouble to give life to new salmon, then die a few days later.

HOW CAN WE TELL WHAT MINERAL A PLANT LIKES?

If you will dissolve the mineral in a liquid and make it radioactive by holding it in the presence of a radioactive substance, the clicking of the Geiger counter in the hands of a scientist will tell him the amount of mineral the plant has taken in or absorbed. This will tell him how much the plant likes the mineral, and also which part of the plant absorbed the most.

WHAT IS TEMPERED GLASS?

Tempered glass is made to be especially unbreakable and strong under certain conditions. This is done by a special heating process and an especially long cooling period. Therefore the glass has additional strength. A good example would be the bulletproof glass used in many banks and commercial establishments for protection. Another example is a glass called Herculite. It is said to be so strong that it could resist the fall of a bowling ball upon it from a considerable height and come out unscathed and unbroken.

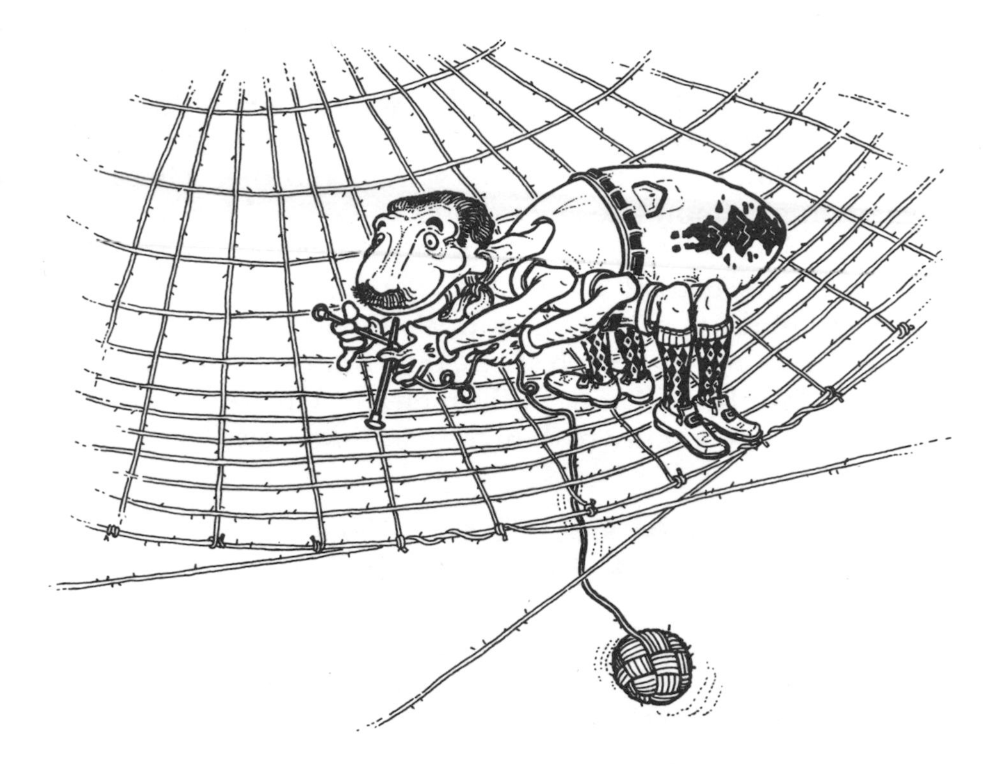

HOW DO SPIDERS MAKE A WEB?

At the hind end of the abdomen of the spider are three pairs of glands, each with a separate tube to the outside. These tubes are called spinnerets. By holding all the tubes together it can make one thick thread, or it can hold them apart for a band of fine separate threads. When a typical spider starts to make a web, it presses the spinnerets against some object, usually high off the ground, and forces out some of the liquid silk. Then it moves away and draws out the sticky liquid which hardens in the air. It climbs down this thread and attaches the other end to some object. It will do this till it forms a figure like the spokes of a wheel. When the spokes are finished, the spider starts from the center and connects them all with a concentric pattern growing larger and larger as it goes round and round. The web will usually hang vertically in order to catch flying insects.

The largest spider web ever seen was about twelve feet in diameter, in Africa. It has been estimated that a spider's silk is so strong that a woven thread one inch in diameter could hold up 74 tons. It is three times stronger than a similar rope made of iron.

Are you wondering why the spider doesn't get caught in its own web? Well, a famous scientist discovered the reason. The spider coats its own legs with an oily substance from its mouth, so that it can pass over the strands without being caught.

DO ANY MALES EVER GIVE BIRTH TO YOUNG?

If your neighbor rushed in and told you that her tomcat gave birth to kittens, your hat would spin around your head in amazement. Well, here are two little fish, the males of which carry the unborn and give birth to

them, while, I presume, the females swim back and forth in great anxiety in the imaginary waiting room of the sea. The females deposit the eggs into the pouches of the male, wherein they become fertilized by the sperm and begin to incubate. The young are born alive, and one observer saw 294 half-inch babies born to a male seahorse in less than twenty minutes.

The seahorse, of which there are 24 different species, and its cousin the pipefish, live in shallow seagrass beds where they can hide. Seahorses are about one and a half inches long, swim upright and look like tiny horses without legs. Their relatives, about six inches long, swim in deeper waters. On the seahorse the pectoral fins are like little propellers that flutter about seventy times per second, while the dorsal fin acts like a rudder and beats in synchronization.

Many of the tiny seahorses are devoured by other fish and only a small percent reach maturity. In fact their own papa will eat them up soon after he gives birth to them, if they should innocently swim past him. Could be they gave him a rough pregnancy.

WHAT GIVES THE FIREFLY ITS LIGHT?

The light which fireflies and glowworms give off is the result of a chemical reaction within their bodies that results in light without heat. The firefly, which is really a beetle about a half inch long, is one of about 200,000 species in the world.

The light comes from the area on the sides of the stomach where a fatty tissue is located. This tissue contains air tubes and nerves. When the nerves stimulate the air tubes, they give off oxygen which combines with six substances in the fat called luciferin. The result of this mixing gives off a heatless light, greenish-white in color.

The natives of some countries use them in a bottle as a lantern, or put them on their boots to light the path. Although they don't get a steady light, because the beetles flicker on and off, they bring enough of them along to make it suffice.

During the Spanish-American war, Dr. Walter Reed is reputed to have performed an operation in Cuba solely by the light of a bottle full of "cucuyo" moths.

HOW ARE METALS ELECTROPLATED?

Pure silver, copper, gold or chromium is placed into a chemical solution and the cutlery or the metal to be plated is suspended in the solution. An electric current is then allowed to pass through the solution. Slowly the cutlery or metal acquires a coat of the copper, gold, silver or chromium, or whatever is used for the coating.

WHAT ARE PARASOL ANTS?

These ambitious ants, which live in very hot countries, cut fresh leaves into pieces much larger than themselves and carry them over their heads like little green parasols. Where to? To their nest, where these tiny farmers raise mushroomlike fungi, the only food they eat. The bits of leaves are used to fertilize the beds of the growing plants.

When the ants reach the nest, the leaf is chewed into a soft wet mass and placed into a room the size of a football. There they plant bits of fungi already growing in their gardens. Soon the wet mass gets covered with the white threads of the growing plants. To keep the plants from growing too large for the nest, worker ants keep snipping off little pieces. Other ants weed the garden to keep harmful plants out.

When a young queen leaves her nest to start a new colony, she carries a bit of fungi in her mouth to start a fungi garden for her young.

HOW DO ANIMALS HELP MANKIND FIGHT DISEASE?

Horses produce the serums for our tetanus shots and for diphtheria antitoxin.

Cattle and *sheep* supply us with adrenalin and thyroxin. Sheep supply us with insulin from their pancreas and *cows* give us the vaccine for smallpox shots.

Fertilized chicken eggs are used to develop antibodies, and *rabbits* and *frogs* are used for pregnancy tests in women.

Mice, guinea pigs, dogs and *cats,* because they are so available, are used for many scientific experiments and explorations.

The *rhesus monkeys,* who suffer many of our diseases, are much in demand by many pharmaceutical and science laboratories.

Because *mice* and *rats* give birth to fifteen or more sets of young each year, they are excellent subjects in studies where genetics and inheritance factors are involved.

Many animals serve as biological controls for the diseases of domestic plants and animals, such as the snails which feed on orange tree fungus.

HOW DOES PLANT LIFE SURVIVE IN THE DESERT?

If you ever get lost in the desert and you're parched with thirst, find yourself a "barrel cactus" and squeeze yourself two quarts of juice. You'll have to crush the tender pulp first, but I'm sure you won't mind this. It may not taste like a cola drink, but it can save your life.

How come so much juice on a dry desert? Well, these plants are not like other plants. Most plants you are familiar with get their food from the leaves through the process of photosynthesis. The leaves also take in carbon dioxide and give off water. This is not so with desert plants, as their leaves have been modified by evolution into spines on the plant which have little surface area from which to give off water. Thus they can better retain their moisture within their thick stems. This is their secret water storage system.

The roots of the cactus plant also play an important part in its desert life. They are covered with a corklike bark and lie close to the surface of the ground, spread out over a wide area in order to absorb all the moisture possible. A small cactus plant about three feet tall may have roots that reach out ten feet or more. Thus desert plants are able to survive long periods of dryness.

WHAT IS THE PRESSURE AT THE BOTTOM OF THE OCEAN?

Compared to the atmospheric pressure of 14.7 pounds per square inch, this pressure is seven tons (14,000 pounds) per square inch, or almost 1,000 times atmospheric pressure. At 3,000 feet down, the pressure is 1,350 pounds per square inch, where a block of wood can be squeezed into one half its volume. At a depth of 25,000 feet, air will be so compressed that it will be as dense as the water.

WHY IS THE OCEAN SALTY?

Scientists once thought the oceans began as fresh water bodies, but today it is generally believed that the primeval seas were initially salty, having dissolved their salts from the rocks underlying their basins. Frost and erosion of continental rocks added salt to the sea, and this slow process is still going on, as it has for hundreds of millions of years.

HUM-M-M-M-M-M-M-M-M-M-M-M-M-

DOES A HUMMINGBIRD REALLY HUM?

The flapping wings of the hummingbird are so rapid that, like the whirring propeller of a plane, only a misty outline can be seen. It is this rapid flapping of the wings that gives the bird the humming sound. Actually its wings flap about forty to fifty times each second, and it has been estimated that its speed in flight reaches about 700 miles per hour. An insect in flight doesn't have a ghost of a chance of escaping the speed and agility of this tiny bird that is only two inches long and weighs no more than a copper penny. To reach deeply into the flowers for nectar, it uses its long tubelike tongue, and like a helicopter, it can fly straight up into the air and remain suspended in front of the blossom while sipping it up.

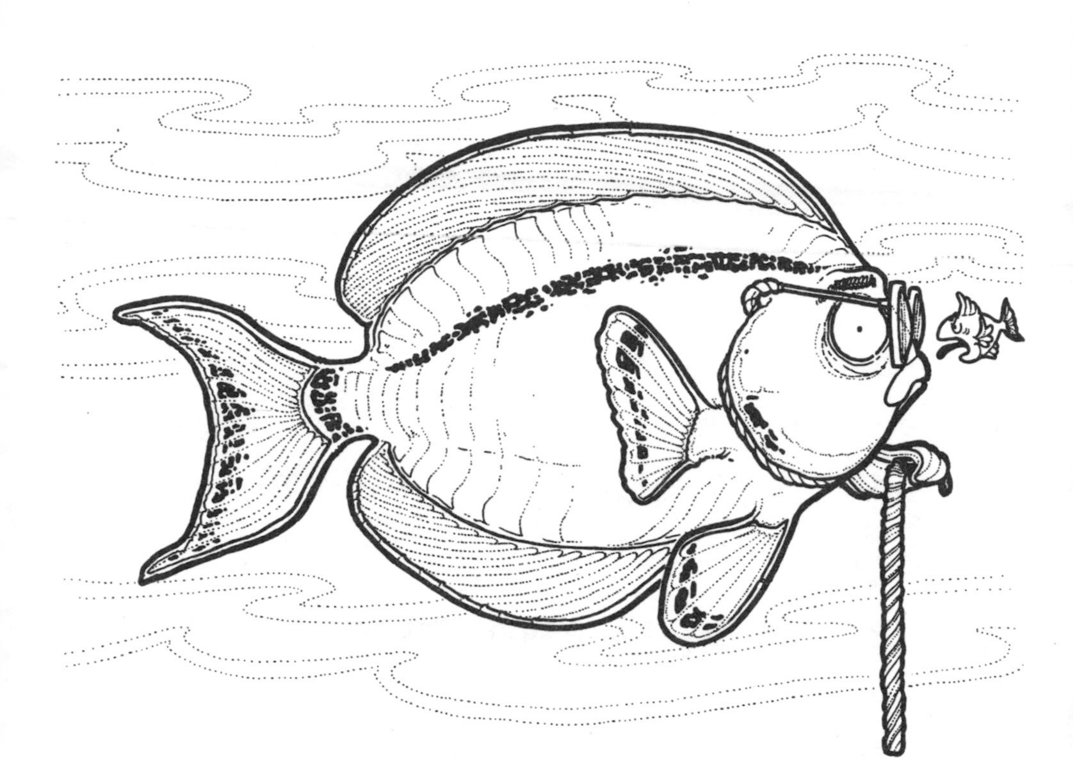

ARE THERE ANY BLIND FISH?

If you hate the sight of cooked fish eyes when served to you, you may now be consoled. There are over seventeen different kinds of blindfish that inhabit the dark waters of caves in the eastern part of the United States, such as Mammoth Cave, Kentucky, and in cave waters throughout the world. These fishes are called blindfish because they are either born with no eyes at all, or with eyes that function for a period and then disappear. However, nature has compensated them for their lack of vision by supplying them with rows of small projections on their heads and bodies to give them an extraordinary sense of touch.

IS THE SUN A SOLID OR A GAS?

By closely studying the sunspots on the sun, it can be concluded that the sun is a mass of gas and not a solid. Sunspots in the upper portion and equator regions appear and disappear, then appear and disappear again. This shows that the sun is rotating. However, the sunspots on the sun's equator rotate faster than the ones at the pole. This shows that the sun is not a solid. If the sun were solid, both spots would appear in the same place as the sun rotates.

HOW FAST DOES A WHALE GROW?

Whales show the most spectacular growth known for any animal. It takes eleven months for a whale to develop from a microscopic egg to a weight of fifteen tons and a length of twenty-three feet before it is born. Then for the next four months the calf continues to nurse from mama whale, after which it seeks its own food. By the end of the first year it is 65 feet long and weighs 65 tons, an average gain of 2.3 tons per month. All this by feeding only on small two-inch crustaceans known as krill.

HOW DOES LIFE GET TO A NEWLY FORMED ISLAND?

If you should ever get stranded on a newly formed island—for example, one of the Hawaiian Islands—you need have no worry at all if you have the patience to wait. After all, what else have you on your hands but time? Just remember that there are plants especially well adapted for dispersal across a sea barrier.

Coconuts floating about in the sea for months will be cast up on the beach and begin to sprout. See? All you need is patience. Fern spores drift in the air like dust, and unless you're allergic to fern spores, you'll welcome them.

Some birds have reached the most isolated of all the Hawaiian Islands, so if your luck holds out, some will soon reach you. Ballooning spiders may also reach you eventually, but I doubt if you'll welcome them.

Seals and penguins can swim vast distances to reach islands, but they will not reach you in the Hawaiians, so don't wait for them. Certain lizards lay their eggs in driftwood that most likely will be cast up on shore. I'm sure you will begin to like lizard eggs.

Certain small lizards, rodents and invertebrates hitchhike rides on certain debris like logs and storm-rooted trees. You may get to like them too. Other animals can swim for many hours in the ocean, and may live long enough to reach a distant island. Who knows? It may be the very newly formed island that you are stranded on.

WHAT ARE THE SUTURES IN OUR SKULLS?

Our skull is made up of eight cranial and fourteen facial bones. The jagged threadlike lines are the immovable bone joints formed during embryonic development, and are called sutures. If you asked any pre-med student why he was using the skull of a cat, dog, sheep, lynx, monkey or any mammal, he would tell you that he was learning all about the bones of the human skull. Why the skulls of these mammals? Because for the most part the same bones, although different in shape and size, are divided by similar sutures in both the human skull and those of most mammals.

These sutures have the characteristic of meandering and weaving back and forth in order to hold the different sections more strongly together.

When we are born, some of the separate bones of the skull can be moved with relation to each other because the joints are of cartilage. This makes it easier for a baby's head to pass through the mother's birth canal at the time of birth.

For some time after birth the top of the skull remains cartilage, and only gradually do the bones fuse together to make a rigid cage or box. When we reach about age 40, only three-fourths of the bones of our body are fully fused.

HOW DOES A BEE MAKE WAX?

Of the three types of bees—the queen, the drone and the worker—the only one that makes wax is the worker bee. How this is done is still a mystery to scientists, but when it is made and leaves the body, it collects on scales known as wax plates found on the abdomen. With this wax the worker bees cleverly build the beehive by instinct into perfectly hexagonal cells, into which the eggs are laid and from which the larvae will mature and come out.

WHICH BODY CELLS DO NOT REGENERATE?

Brain cells once destroyed by injury or lack of oxygen will not regenerate. If the damage is serious, a person may be left deaf, blind, speechless or paralyzed for life, depending on the part of the brain injured. From the moment of birth the average human being loses brain cells. They die at a rate as high as 100,000 per day by age 60, and, unlike other body cells, they are not replaced. Dr. John Kangas, Director of the University of Santa Clara Counseling Center, believes that despite the diminishing number of brain cells, the I.Q. may actually increase with age.

HOW DOES A BIRD LOCATE A WORM IN THE GROUND?

Before the sun rises to heat the surface of the earth, the worms usually crawl up while the earth is still cool and damp with the morning dew. Unfortunately, this is the time they fall prey to the hungry early birds. When a bird stands on the ground near a worm that is crawling underneath, it can feel the earth's vibrations with its very sensitive feet. It will also cock its little head to put into operation the low-frequency apparatus of its ears. Then, when it has zeroed in on the victim, it pierces the earth with a sudden stab of its beak, grabs the worm and pulls it out.

WHY DOES A MOSQUITO BITE ITCH AND SWELL?

When a burglar goes to a job, he takes with him the best tools—especially if he has to make a hole in a wall or a safe. Well, the mosquito does likewise. To make a hole in your skin and burgle your blood, she carries with her the best tools.

The first important tool is her beak or proboscis. This is the part that she injects into your skin. It is made up of four parts, one of which is a channel through which she can pour her saliva into your wound. The saliva is her second important tool because it acts as an anticoagulant. It dilutes the blood so that it can be sucked more easily and flow more smoothly while she siphons it off. The injected saliva produces a mild allergic reaction in your body at the site of the wound, which results in itching and swelling. So swift a burglar is the mosquito at her job that the whole act of assault, breaking and entering, injection of saliva and stealing the blood, all takes about the blink of an eye.

Incidentally, the beak of the male mosquito is not strong enough to pierce the skin. It is only the female mosquito that does the biting.

MUST PEOPLE BE TONE DEAF?

The answer is no. There is no such thing as tone deafness. With proper training and effort, one should not be deaf to any sound at all. But if a person is hard of hearing, that is a different story. This is a form of hearing deficiency which renders all sound, including melodic tones, hard to understand clearly. If, however, a person has no hearing deficiency and keeps saying he cannot carry a tune, you may be sure he was too lazy to train himself correctly.

WHY ARE MINER BEES LIKE CONDOMINIUM DWELLERS?

The miner bees are one of a group of solitary bees where males and females live in small apartments of their own, as distinguished from the social honeybees who live in a hive. These bees dig a tunnel in sandy banks to raise their brood. Then the female digs her own compartment, a little off the main trunk line, to lay her eggs. The entire nest, which consists of the main tunnel and its hundreds of separate "apartments," contains thousands of bees. All it lacks to compare it to a modern condominium is a swimming pool and a shuffleboard. But it does have a security guard at the entrance to permit a known dweller to enter (it can tell by the smell) and to fight to the death to keep a stranger out.

HOW DOES A BEE KNOW WHERE TO MAKE A "BEELINE" TO NECTAR?

Whenever a bee returns from a feast of nectar, his friends soon fly to the same spot without being led by him. How did he tell them? If you say that smell helped, you may be right if the distance is short. But if it is over 3½ miles from the hive, it would take more than smell. A German professor who made a study noticed that a returning bee did two kinds of dances in the hive. One, called a "round" dance, in which the bee made a number of turns, told them the distance; the other, a "waggle" dance, where the bee made straight runs back and forth while waggling, gave them the direction. The experiments took years and were quite conclusive. They involved the time of day and the position of the sun, because it pointed out the angle between the nectar, the hive and the sun. Thus a tiny bee can tell its fellow bees the distance and direction of the nectar without a blackboard, ruler or piece of chalk.

WHAT ELSE BESIDES FISH DOES THE SEA GIVE US?

Sulfur is being tapped by drilling undersea deposits from platforms in the Gulf of Mexico.

Oil is being recovered from beneath the ocean floor to the point where about 16% of the world's oil supply was obtained from this source by 1965.

Phosphorite. Nodules off the California coast give California an abundant supply of phosphate fertilizer.

Diamonds, pearls, sand, gravel and shells come from the sea floor or beaches.

Ores of *tin, thorium* and *titanium* come from the floor of the sea.

Bromine. 75% of the United States supply is extracted from the sea.

Magnesium. All of the United States supply comes from the sea.

Salt. Very much of this comes from the sea.

Manganese. 1.5 trillion tons of manganese nodules on the bottom of the Pacific Ocean contain 50% of manganese and small amounts of *nickel, copper, cobalt* and other metals.

HOW MANY SENSES DO WE REALLY HAVE?

Most people think of our five traditional senses as the only ones we possess, but in reality there are about twelve. The other seven senses are more or less discussed in textbooks, but for some reason have never been entirely included in a complete list. Here is a list of all twelve senses.

1. Sense of taste
2. Sense of smell
3. Sense of hearing
4. Sense of touch
5. Sense of sight
6. Sense of pain
7. Sense of warmth
8. Sense of cold
9. Kinesthetic sense (deals with muscle expansion and contraction)
10. Sense of hunger
11. Sense of thirst
12. Sense of equilibrium or balance (A person who has had a leg or arm amputated must learn to walk in a balanced position all over again, because the amputation brings about an imbalance)

WHY CAN ANIMALS SURVIVE IN THE DESERT?

Well, we know for sure that they don't pray for rain, they don't go out seeding clouds, they don't do Indian rain dances, and they don't ever worry whether rain clouds show up or not. Why? Because they can very well do without water in their mode of life or in their physical structure.

Let us take, for example, the kangaroo rat. It is a small desert animal of western North America that gets its name because it looks like a combination of a rat and a midget kangaroo. In reality it is neither. By using its long hind legs and its tail like a kangaroo, it can make extraordinary leaps of over six feet over desert ground. To conserve its body water content, it stays in its underground burrow during the daytime and comes out only at night. The amazing part is that it doesn't drink any water at all. All the dry food it eats turns into water in its body. By means of its special kidneys, the water content in its body is recycled. This, of course, prevents a water loss.

Except for some minor differences, all that was said about the kangaroo rat can be said about most of the other desert animals, such as the pocket mouse, the antelope ground squirrel, and the jerboa.

WHY DOES FUR BRISTLE ON A FRIGHTENED CAT?

Just like human hair, fur has tiny follicles at its base in the skin. From these follicles, tiny muscles run out in a slanting direction. When a person is suddenly frightened, there is an involuntary adrenaline reaction. But with cats, a sudden fright results in an involuntary reaction that makes the tiny muscles around the follicle contract. This makes the fur stand up straight.

It is a clever defense mechanism which makes the cat look almost twice its normal size. Not only does the fur stand up straight, but the cat also arches its back and raises its haunches as high as possible to simulate power and height. If the foe still challenges, the cat will heighten the intimidation by hissing and clearing its throat in a rasping fashion as if it were about to expectorate in the face of its foe. If all this does not help, it will turn its back and run like the devil up the nearest pole or tree.

DO SOME PLANTS EAT INSECTS?

If it surprised you to hear that a man bit a dog, or a chicken ate a fox, how about plants that trap, kill and eat insects? To do this, nature has provided certain plants with a device to trap the unwary insect. The bladder of the *bladderwort plant* drowns its victim after the victim is caught by the plant's valvelike trap door. The *pitcher plant* has tube-shaped leaves that hold rainwater. Insects caught here are also drowned, and in both cases are later digested by the plants at leisure.

Sundews have rosettes of leaves with sticky hairs. When the insect is caught by these hairs, the leaf margins quickly curl around it and proceed to enjoy the contents of the succulent but unwilling donor.

Venus's flytrap has leaves that work like a steel trap. The leaves have a pair of lobes, on the surface of which are three sensitive hairs. When an insect alights on one of these hairs, it is the last that it is going to see of this world. The pair of lobes close like a trap, holding the insect inside. Special glands of the leaf secrete a fluid that helps the plant to digest the soft parts of the insect. When the victim is nothing but a skeleton, the trap opens and the innocent-looking leaves are in position again for another hapless victim.

WHAT IS MEANT BY 20/20 VISION?

When you tell your eye doctor that either your eyes are bad or your arms are too short when you read a newspaper, he knows what you're talking about. But after he examines you, and tells you that your right eye has 20/20 vision and your left eye has 20/30, do you really understand him?

The figures 20/20 mean that at a distance of twenty feet, you can read letters of a normal size for that distance. If, for instance, you had 20/30 vision, it would mean that you could read letters from a chart at twenty feet that someone with normal vision could read at thirty feet. In most cases, both eyes vary in capability. One eye may have a 20/20 capability, while the other may have a 20/30 or 20/40 capability. Seeing at a distance remains normal much longer than seeing at close range or reading a newspaper. One or both eyes may see normally at a distance, but even so, one or both eyes may need correction with lenses.

HOW DO FRUITS AND VEGETABLES GET THEIR COLOR?

One day a little girl was seen pouring blue ink-water on her tomato plants in her small garden. When her surprised mother asked her what she was doing, she replied, "I want to see if the blue water will grow blue tomatoes."

She thought this was a clever idea for an experiment, and she wasn't so wrong. Some day, with her inquisitive mind, when she is old enough to understand the laws of heredity, she will know why all the colored water in the world could not change the color of the red tomatoes, green plums, or the gold of lemons or oranges.

The wonder of the phenomenon, of course, is how the dissolved minerals in the ground, being colorless, turn into different colors when absorbed by the roots of a red tomato plant, or a red cherry tree, or an orange tree. The secret lies in the genes of the chromosomes of the individual seeds of the different plants, flowers, fruits and vegetables. The factors for color in all of them were predetermined, just as were predetermined the factors for color in our own genes, for our hair, our eyes and our skin.

HOW OLD IS THE ORANGE SOIL BROUGHT FROM THE MOON?

The orange soil brought back from the moon by our American astronauts looked remarkably fresh, and because the craters resembled the volcanic vents that we have here on earth, it was first believed that volcanic activity occurred on the moon as recently as 200,000 or 300,000 years ago, and spewed forth this orange soil. But by using the "atomic clock" dating techniques, Dr. Oliver Schaeffer and his lunar analysis team at State University of New York at Stony Brook determined that the material was 3.71 billion years old, and within the age range of other moon samples brought back to earth by them.

DO STARS REALLY TWINKLE?

All the shining stars we see at night are really like our sun in that they are balls of flaming gases sending out rays of light in all directions from billions of miles in outer space. The light rays that reach our earth have to pass through different layers of air in our atmosphere before they can be seen by us. The light waves from the edges of the beam travel through different thicknesses of air than the center waves of the beam. This causes an irregular arrival of the light waves, which gives the viewer an impression of twinkling.

HOW IMPORTANT IS THE PROCESS OF OSMOSIS?

Osmosis is the process by which a dissolved substance in the form of a liquid or a gas passes through a membrane from one medium to another.

If it were not for this process, our vegetable world could not exist. It is through this process that dissolved minerals underground pass through the thin membranes of the tiny root hairs in order to get into the plant. Thus the plant receives its nourishment so that we may have life and nourishment too.

Another example of this vital process is concerned with our body's supply of oxygen. The oxygen which we breathe into our lungs passes through the very thin membranes of the tiny capillaries, which then carry it to all parts of the body by way of the red corpuscles in the bloodstream. In the same way, carbon dioxide leaves the bloodstream.

By this process also, water is absorbed from our stomach and ends up in the bladder.

Now you can see how important this process is to our lives.

CAN AN ELECTRIC EEL REALLY SHOCK?

The so-called electric eel is really not an eel, but a greenish-black fish two to four feet long. It has a flat head with its eyes far towards its mouth. The inner organs of this fish are located in the first one-fifth of its body, but the other four-fifths of its body mainly contain the organs which make the electric current. There are three pairs of organs that make the electric current, which flows from head to tail.

If you don't think they make real electric current, consider the fact that experiments have shown they can generate enough electricity in their bodies to knock down a full-grown horse or drive a small motor. A large electric eel can discharge up to 600 volts of electricity.

DO PEOPLE EAT SEAWEEDS?

For many years seaweeds have been raised in Japan for food. Even our own health stores now carry some form of seaweed. Dulce and kelp are two very well known and popular items sold in the United States. Kelp grows off the coasts of Ireland, Scotland and Brittany. Large masses of giant kelp grow off the Pacific coasts of North and South America. Partially dried kelp plants are used as fertilizer. Kelp ash is an important source of iodine, giving at times as much as 15 pounds per ton. It also contains salt, sodium carbonate, sodium and potassium sulfates and potassium or sodium iodide. Great quantities of seaweed are found in the Sargasso Sea. It is food for livestock, used in the making of bread, candy, canned meat, ice cream and jellies.

DO FISHES EVER GO TO SLEEP?

Many fishes in the coral reefs do sleep, and actually assume sleeping postures, typically by standing on their tails or leaning on a rock. However, a majority of fishes are in continual motion, but do have times of lesser activities.

WOULD YOU BELIEVE A FISH HAS LANTERNS?

Both the lantern fish and the deep sea anglers have a luminous gland system in their bodies. The organs or glands on the lantern fish that give off light look like little pearls. They have several series of lighted organs on the sides of the head and body to guide them as they swim about in the dark depths of the ocean where they dwell. Most marine fishes that light up are red in color.

CAN A FISH BUILD A NEST?

You may have heard of walking fish, flying fish, tree-climbing fish, and now you will hear about nest-building fish of the northern hemisphere. The stickleback is the only known nest builder among the fish. It gets its name because instead of scales and fins it has hard plates that are more like sharply separated spines.

These fresh- and salt-water fish, from one to seven inches long, can build a nest in the shape of a muff by using a glandular cement which they manufacture themselves. For nest material anything will do, such as sticks, roots and fibers of water plants. When the nest is finished, the female lays the eggs while the male stands guard outside. After the eggs are hatched, he also watches them for several days.

But that is as far as their affection goes. They are known as a fighting, greedy lot, ready to eat the young of other fish and bite the fins off their peace-loving friends when placed in an aquarium.

WHAT FRUITS WERE UNKNOWN TO OUR FOREFATHERS?

Loganberry is a purplish-red form of the blackberry. It was first introduced in Santa Cruz, California, in 1881.

Avocado is a green round or egg-shaped fruit that can weigh from half a pound to three pounds. It is grown chiefly in California.

Persimmon. There are three kinds: the oriental, the lotus, and the American. They first came to the U.S. in 1825.

Guava is a yellow pear-shaped fruit as big as a hen's egg. Believed to have come from Brazil, it now grows in Florida and California.

Banana. Its original home was Asia and it is now grown in Central America, Mexico, West Indies, Ecuador and Brazil.

Kumquat, like a tiny orange, is a native of China and cultivated in Japan. It is now grown in Florida and California.

Grapefruit may have originated in the West Indies, and was brought into Florida in 1820. Now it's grown there and in California.

Papaya is a large, mildly sweet fruit that could weigh 10 to 12 pounds. It originally came from Central America.

Mango. A fruit with a juicy yellow pulp, originally grew wild in Southeast Asia. It was brought to the United States in 1700.

Casaba melon, called a winter melon, came to California from Turkey in 1871.

Tangelo, which is a cross between a grapefruit and a tangerine.

WHO HAS BEEN HERE LONGER, THE COCKROACH OR THE DRAGONFLY?

The roach was on this planet about 280 million years ago, and the dragonfly about 320 million years ago. They are what you really can call old-timers. There are about 3,000 species of roaches in the world, and they can live almost anywhere except in extremely cold regions. Their diet includes any food that man eats and also clothing and wallpaper. You probably know they are one of the fastest runners if you ever tried to smack one with a slipper. Generally the roach is considered a pest, but its ego spells its doom. Roaches like to groom themselves by scraping foreign particles and dust from

their bodies with their legs, then drawing the legs and feelers through the mouth to get rid of the powder. Thus, if we spray the roach's surroundings with insect powder, we will soon find the bug on his back, heading for his favorite crevice in the sky.

On the other hand, the dragonfly is our friend, because flies and mosquitoes are to it what apple pie and ice cream are to us. And like us, the dragonfly never tires of its delicacy. Its bulging eyes can spot an insect 60 feet away, and it can catch the insect with a swoop of 60 miles an hour. When he arranges his six spiny legs in flight in the shape of a woven basket to catch an insect, the object of his gustatory desire doesn't have a chance. It is even devoured while in flight.

HOW DO TERMITES DIGEST THE WOOD THEY EAT?

Symbiosis is the process of two plants or animals living together for their mutual advantage. This situation exists between the termite that chews the wood and swallows it, and the microscopic protozoans that live in its intestine and digest the wood for the termite.

Wood is composed chiefly of cellulose, a substance that very few animals can digest. But the protozoans, who can digest cellulose, create an end-product which the termite uses as food. Thus both the termite and the protozoans benefit by living together.

The only one who does not benefit from this happy symbiotic arrangement is the property owner who finds one day that he has to replace a wooden floor or an upright beam in a wall—thanks to the insects who use his house as a free supermarket.

ARE THERE ANY DRAGONS STILL AROUND?

Not exactly. But there is an East Indian lizard called the dragon of Komodo that could bite the steel pants off an armored knight of the Round Table.

He is ten feet long, weighs about 150 pounds, and gets his name from a small island in the East Indies. Besides looking and being ferocious, his long tail with its dull-colored scales makes him a tough opponent in any fight. He has a wide red mouth, full of rows of razor-sharp teeth. Animal hunter that he is, he is well provided with a keen sense of sight and smell.

With his strong sharp claws he digs a cave for his nightly home, but comes out during the daytime to hunt for prey.

And why could this lizard-dragon toss a knight around? Well, he has been known to tear off the hindquarters of a boar and swallow them, bones and all. A University of Florida zoologist saw a Komodo dragon of medium size bite the head off a 90-pound hog, swallow it whole, and proceed to gobble the rest of the hog in just 17 minutes.

HOW DOES A LIVER PILL FIND THE LIVER?

Let's make believe that you are a liver pill and were swallowed in a hurry to do a merciful job. How will you find your way to the liver? Truth is, you won't have to do a thing. You will be so thoroughly dissolved by the enzymes of the mouth, stomach and small intestine that your hard crust will totally disintegrate into ultramicroscopic molecules. You will then float around in the blood stream, awaiting your destination, the liver. You will eventually get there because no other organ cares for you or wants you. This is true of kidney, heart, pancreas, stomach or any other pills.

For some unknown reason, the liver has a special attraction for your molecules which no other organ has. That's why they call you a liver pill, because your molecules have a beneficial effect upon the liver and no other organ in the body. Only after years of experimentation was it found that your molecules benefit the liver and do not harm the other organs. That means there were no after- or side-effects after you were absorbed by the liver.

The Food and Drug Administration oversees the proper production of safe drugs and foods, and gave the "go-ahead sign" to your company to manufacture you, because you were safe and helpful, and knew how to behave yourself by not harming the other organs.

DOES WATER DEFY GRAVITY TO REACH THE TREETOP?

Wouldn't it surprise you to see water run uphill, when you know that water or any liquid only flows downhill? Well, water goes up from under the ground to the tops of the tallest trees, some of which are over three hundred feet high, which is as high as a thirty-story building. How does the water get up there?

The first guess might be capillarity or capillary action. Blotters absorb ink and water because of capillarity. In a glass tube with water in it, you can see the water rise and cling to the sides of the tube. But this cannot explain how the water rises 300 feet. We have experiments to prove this. However, the pressure of the air, which is 14.7 pounds per square inch on all surfaces, can only hold up a column of water up to thirty feet. How about the other 270 feet? What force brings the water up there?

The answer could be transpiration, a word which means the breathing of leaves. When a leaf gives off water from its surface, this is part of transpiration. When this takes place, every molecule of water that leaves the leaf pulls up another molecule to the top of the leaf. This happens because all molecules in liquids and solids are tied to each other by a strong force. Thus, with thousands of leaves on a tree, each giving off billions of molecules of water by evaporation, it is believed that all three actions have a hand in the process. Thus, air pressure, capillary action, and the transpiration or breathing of the leaves makes the water rise to the top. This is the best theory we have today about this process.

HOW DOES OUR NOSE KNOW?

If you were an old-timer who lived in a four-story walkup, you would probably smell Irish stew on the first floor, Italian lasagna on the second, Jewish chicken soup on the third, and on the fourth, where you live, no smell at all. The reason? Your wife wasn't cooking. She was opening some kind of can for dinner. But how did you know what your neighbors were cooking? Well, basically, molecules are always in motion, and those delicious-smelling molecules passed through your nose and struck a filmy mucus that covers the olfactory tissue in the back upper region of your nose. Here bathe thousands of tiny wavy hairs that help to filter and warm the incoming aromas that contain those delectable molecules. When those tiny hairs get a whiff of the delectable aromas, they are immediately stimulated. And who can blame them? Thus stimulated, they activate the olfactory bulb. You can almost hear those little cells saying, "Wish we were invited to that dinner." The bulb, now activated and excited, immediately contacts the brain through the olfactory nerve cells.

In order that you may not miss out on any particular smell, nature has provided you with the capacity to detect four primary smells or aromas. They are: fragrant, acid, rancid, and burnt. From those four primaries comes the blending of every aroma you could smell.

No sense of smell could ever beat that of the emperor moth, who can detect a female moth about seven miles away.

WHAT PRODUCES ANNUAL RINGS IN TREES?

Annual rings found in the trunks of trees and their branches are the result of varying climatic conditions during different periods of the year. When water is plentiful, as in the early spring, thin-walled cells which contain a larger central cavity are produced by the cambium. In the summer and fall, a time of less

rain, the tracheid cells which are formed have thicker walls and smaller cavities. Thus it is that good years produce very thick annual rings and drought years produce very small rings. But it is the alternation of spring and summer wood that creates the lines called annual rings, dividing the two distinctive types of cells.

HOW DOES YEAST MAKE THE DOUGH RISE?

Everyone knows what yeast does, but few know that it is a mass of tiny one-celled plants. Like mushrooms, they belong to a group of plants called fungi. Yeast plants multiply rapidly, either by budding or by fission. In the budding process they grow and pinch themselves off from the main plant, and by fission, they simply divide in two. When yeasts grow, they form two enzymes called zymase and invertase. After the yeast is mixed with flour and water in proper proportions, it is kneaded thoroughly, covered over and set aside to rise. The mixing of the yeast with the dough to ferment is called leavening the dough. The enzymes from the yeast cells attack the starch in the flour and change it to suger. The sugar is then changed to alcohol and carbon dioxide gas. The gas bubbles up through the mixture, which makes the dough light and porous. When the bread is baked, the alcohol evaporates and the yeast plants are destroyed. So if you want your morning biscuits light and fluffy, don't slam doors while the bubbles are rising in the dough.

WHY DOES A SIPHON DEFY GRAVITY?

We know that water does not flow uphill because that would defy the law of gravity. Yet that is exactly what a siphon does when it operates. Water flows up into the bent tube from a container holding the liquid, then flows across the tube and down again into another container held at a level below the surface of the first container. These are the exact conditions of a siphon except that to start the siphon to flow, the tube must be full of the liquid to begin with.

The water flows up because two unseen forces are lending assistance, namely liquid and atmospheric pressures. Since the up-flow tube is shorter than the down-flow tube, the uneven liquid pressure is greater in the outward direction. Even though the air pressure is the same in both directions at the start (14.7 lbs. per square inch at sea level), as soon as the flow starts due to the difference in liquid pressures, the atmospheric pressure on the surface of the liquid in the container forces more liquid up through the short arm in order to prevent the formation of a vacuum.

HOW DO WE SEE?

Let us say you are looking at an American flag on a flagpole. The image of the flag passes through a thin layer of tissue called the cornea, then through a water-like liquid, then through the opening called the pupil, and then through the lens. The image then passes through another liquid and finally lands upside down on a dark tissue called the retina. This tissue lines the back of the inside of the eye and could be compared to the film in a camera. From the retina the image is carried to the brain by the optic nerve, the nerve of sight.

If you put the retina under the microscope, you would see two kinds of cells. Some look like rods or little pencils, and some like cones. Each one does a special job. The rod cells tell your brain how much light is on the flag, and the cone cells tell your brain what colors are in the flag. These two things, the light and the colors, excite the nerve cells, which pass the message along to the brain through the optic nerve.

But just how the brain receives the image that it can store away in your mind, to be seen again in your mind's eye 20, 30, or 40 years later, is one of the miracles of nature that to this day remains a mystery to science.

HOW CAN A MAN WITH A PULLEY RAISE A PIANO?

A pulley is a grooved wheel rotating on an axle. It may consist of one or more wheels, all of which will turn by the friction created when a rope is pulled over the wheels. If a movable pulley is used with two cords, supporting a 200-pound piano, the force needed to raise the piano is only 100 pounds, or just half. If four ropes and two pulleys are used, then only 50 pounds of force is needed to raise the piano. If eight cords went through the pulleys, then only 25 pounds of force would raise the 200-pound piano. What you save in power you expend in the amount of rope pulled through the pulleys. When you use one pulley and two cords, you pull two feet of cord for every one foot of lift. When four cords were used, you had to pull four feet for every foot of height. But since you wanted to lift the piano without too much effort, you didn't care how much cord you pulled. That is why a pulley is called a simple machine. A man can do a great amount of work with very little effort.

HOW DOES SCIENCE EXPLAIN MAGNETISM?

Can you imagine something taking place right before your eyes, and not being able to explain it? Well, that's what happens when scientists see a magnet lift a piece of iron or steel. They cannot explain the phenomenon.

Of course, we have a theory. We know that magnets have a north and a south pole, and that like poles repel each other while opposite poles attract each other. There are other things that magnets do, but how they do them is not known. The best answer we have is that magnetism is related to electricity. We know that a force is there, because we can see what it does, but we can neither see nor explain the force.

Now you can be proud. You know as much about magnetism as all our scientists do.

HOW DOES THE FOOTBRAKE STOP A CAR IN MOTION?

Under your brake pedal is a hydraulic master cylinder which contains brake fluid. Pressing on the brake forces fluid through the lines till it reaches a brake cylinder in each wheel. Each wheel cylinder has two pistons that expand when the fluid pressure hits the cylinder. The expansion of the pistons brings pressure on the brake lining which in turn pressures the drum attached to the wheel.

About 1653 a French scientist named Pascal stated the law of pressure which applies to a fluid or a gas. It states that pressure applied to an enclosed fluid is transmitted equally in all directions without loss, and acts with equal force on equal surfaces.

This means that if you use a small pressure of 10 pounds per square inch on your brake pedal, the hydraulic fluid is forced through the tubes carrying it, to the brake linings whose square inch area may run from 60 to 120 inches or more. Thus, if each square inch of lining receives a pressure of 10 pounds, and your brake lining has an area of about 80 square inches, each wheel receives a pressure of 800 pounds, giving a grand total of 3,200 pounds of pressure for the four wheels.

WHAT ARE FIRST-, SECOND- AND THIRD-DEGREE BURNS?

Burns are always serious because of the danger of infection while the damaged tissues are healing. In a first-degree burn, no skin is broken, but it is red and painful. In a second-degree burn, the burned area develops blisters and is very painful. One must not try to open the blisters. In a third-degree burn, both the outer layer of the skin and the lower layer of flesh have been burned. This is the most serious of the three types, as the possibility of infection is greatest. Medical attention is necessary.

WHAT ARE SIMPLE AND COMPOUND FRACTURES?

A fracture is the breaking of a bone; the two types are simple and compound. In the simple fracture, the bone has been broken but the skin has not been pierced. In the compound fracture, both the bone and the skin have been broken. Both cases must be treated carefully and get medical attention. Of course, both are serious because the simple fracture could easily be turned into a compound one if care is not taken, and the compound fracture may be open to infection, so that needs special care.

HOW OLD IS TOPSOIL?

Soils throughout the world vary in quality, and the quality depends upon its content and the percentage of various elements it contains. Soil that is ideal for oranges in Florida may not be good in Oregon where the climate is much different. Each fruit and vegetable to be ideally grown needs proper soil. But as you can now see, what is proper depends upon the location and the climate.

Soil generally is a mixture of sand, clay and humus,

which is the remains of decayed plant and animal life. Underneath the layer of topsoil is a layer of subsoil two or three feet deep, and directly below this, a layer of rock.

The best type of soil is topsoil. Scientists estimate that it takes between 500 and 600 years to make a one-inch layer. Therefore, a six-inch-deep layer would have begun to go through the natural processes of becoming topsoil some 3,600 years ago, or about 1,600 years before the birth of Christ.

No wonder the farmers of the Dust Bowl left their farms when drought and winds blew all their topsoil away.

HOW HAS THE ELECTRON MICROSCOPE HELPED SCIENCE?

In an ordinary optical microscope, an electric bulb or sunlight is used as light for the stage. But the electron microscope uses streams of moving invisible electrons shot from an electron gun to supply the "light." Electromagnets are used to refract and magnify the images.

With the electron microscope magnification of as much as 30,000 diameters can be obtained. But the results do not end there. The enlarged photographic images can then be further magnified by ordinary optical means. Thus a final magnification power of 500,000 or more diameters is now achieved.

Today, objects such as viruses, which were among the minutest and most elusive of organisms to have escaped discovery, can now be brought within view and studied in minute detail. To add further credit to this most amazing instrument, a picture of atoms was taken when it photographed the pointed end of a needle.

How would you like to see a picture of your nose taken with this microscope, enlarged 500,000 times its real size?

WHY IS THE SKY BLUE?

The atmosphere of the earth is about 100 miles thick, and is composed of many gases. Some of them are hydrogen, oxygen, nitrogen, argon, krypton, neon and helium. When sunlight passes through these gases, the different rays are scattered or diffused. The seven rainbow colors and all the intermediary blended colors that white sunlight breaks down into are diffused through the gases of the atmosphere, including the foreign substances in it. But since the blue, purple and violet rays are much shorter than the red rays, they are diffused many times more than the red rays and therefore dominate the sky.

HOW DO PILOTS KNOW HOW HIGH THEY ARE?

In every plane that flies there is an altimeter of some kind. This is an instrument that tells a pilot his altitude. There are several kinds. One is based on the principle of barometric pressure, and makes use of the fact that the atmospheric pressure varies with height. This is not always accurate due to differences in air pressure caused by weather.

Another altimeter makes use of the radar principle, or the high frequency radio echo. With this instrument the pilot sends down a radio beam that tells him the height by the time it takes for the beam to hit the ground and return. Since the speed per second with which the beam travels is known, the time it takes for it to return, divided by half, gives the pilot the height. Of course, the pilot has no real figuring to do, as the instrument is graded and marked and it just takes a glance of the eye to see.

HOW DOES LIQUOR AFFECT A DRIVER?

Alcohol acts as a narcotic or depressant on the nervous system and the brain. Our nervous system, which sends messages to the brain by nerve cell connections called axons and dendrites, is easily upset by the intake of too much alcohol.

The nerve connection called a synapse takes longer to occur because the nerve endings now bathing in the alcohol in the blood stream are dulled into apathy and inertness by the absorption of the alcohol. Thus, a car being driven by a sober person at the rate of 50 miles per hour should stop within 242.5 feet if the brake is applied promptly in view of a possible accident. The car of a sober driver will have moved 55 feet between the time he spotted the impending danger and the time his foot reached the brake.

But a drunk driver whose reaction time has been slowed by alcohol will have traveled nearly 100 feet or more before he reaches his foot to apply the brake.

Although it takes about 500 bolts and nuts to hold a car together, it only takes one intoxicated nut to scatter a car all over the road.

WHAT IS THE INTERNATIONAL DATE LINE?

If we travel in a westerly direction, say from New York to California, for every 15 degrees longitude we pass, we gain an hour. Conversely, traveling in an easterly direction, an hour is lost for every 15 degrees longitude passed. That is why California time is three hours earlier than New York time. When it is 6 P.M. in New York, it is only 3 P.M. in California. If we keep up the game of saving one hour for each 15 degrees longitude passed, we can be one day ahead or one day behind, depending upon which direction we traveled. So, to maintain uniform calendar dates, one must correct his watch when he passes that imaginary line called the international date line. This line is the 180-degree meridian because it supposedly divides the earth in half going from east to west or west to east. In reality, it is a politically arranged line, zigzag in shape.

COULD THE WORLD EXIST WITHOUT CHLOROPHYLL?

Chlorophyll is one of life's most important substances. Without chlorophyll, plants cannot take the sun's light energy and change it into food or chemical energy. By the process known as photosynthesis, the plant takes carbon dioxide from the air, water and dissolved minerals from the soil, light energy from the sun, and chlorophyll manufactured by the enzymes of the plant, and puts them all together in an unbelievably mysterious way to manufacture sugar, fats and proteins.

But without chlorophyll the plants could not manufacture the food which supplies all the living creatures of our world. Yet the wonder of it all is that, although it is an important part of food-making in the plant, no part of itself is used up or becomes part of the starch. It only helps to do the work, but loses no part of itself. In science a substance that acts that way is called a catalyst.

HOW DID MEDIEVAL PEOPLE TELL TIME?

Thousands of years ago, long before man learned to use watches and clocks, primitive man used primitive means. In those years you might have heard a voice calling, "Hey, Ramstail, get up and catch some grasshoppers. It's near lunchtime." Mrs. R., in her leopard skin and with threatening spear in her hand, was watching the tree shadow near her cave and knew it was noon because the shadow was very small. Of course, once this was known, the sundial was sure to follow.

Then came the hourglass, which used falling grains of sand. It was about three thousand years ago that the Chinese applied the same idea, using water dripping from one container into another, to tell the hours of day and night. When candle wax came into use some two thousand years ago, the measured markings on burning candles told the hours of the day.

It was about the year 1300 that a man named Henry de Vick invented a clock with wheels, a dial and an hour hand, and by 1700 there finally came the clock with a minute and second hand and a pendulum. Well, it was about time.

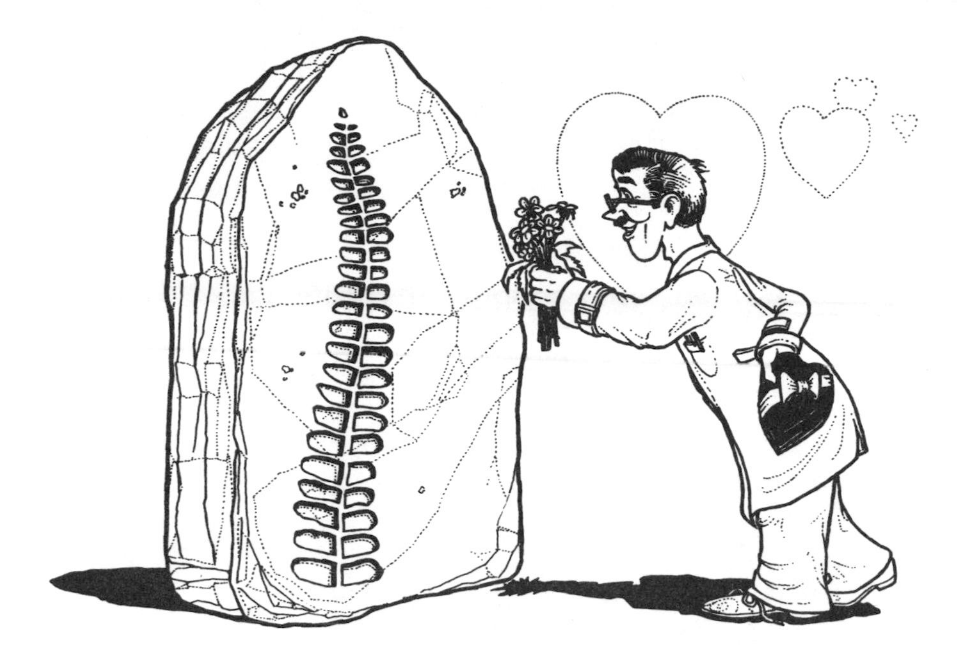

HOW DO SCIENTISTS DATE THE FOSSILS?

There are three methods used to approximate the age of fossils.

1. Chemical dating. Where water contains fluorides or other chemicals, the longer a fossil is exposed to it, the more of the chemical it will contain. The same is true for uranium. The percentage contained approximates its age. A converse method is one that reckons the amount of remaining nitrates in a fossil to approximate its age. Also, when calcium in bone is replaced by silicone, the percentage of remaining calcium will give an approximate age.

2. Radioactive method. Through the ages, potassium degenerates into argon, uranium into lead, and carbon 14 (a radioactive isotope of carbon) degenerates into carbon 12. With a fair knowledge of degeneration periods of elements, a simple percentage determination of the remaining amount of the degenerated substance helps arrive at an approximate date.

3. Stratigraphic method. This method is one of comparison. Fossils to be dated are compared with fossils already found in strata of rock and classified according to the age in which they belong. Besides the strata, other factors must be considered, such as location, climate, history of the surrounding terrain and the condition of the fossil itself.

Only recently it was determined by the amber-colored dust recovered from the moon by our astronauts that the moon became dormant 3.71 billion years ago. The potassium-argon dating method was used.

HOW FAST CAN YOU STOP YOUR CAR?

A scientific table has been compiled that shows how far your car travels after you spot an oncoming car, and the time it takes to step on the brake. Then it shows you the distance your car travels after you step on the brake and the total distance your car traveled till it comes to a complete stop. If your nerves are normal and you do not panic, if the ground is dry and your brakes are in good condition, the figures of this table apply.

miles per hour	Distance between impulse and brake response		Distance after the brake was applied	total distance
20	22′	from the time	30.0′	52.0′
30	33′	you see the	67.5′	100.5′
40	44′	oncoming car	120.0′	164.0′
50	55′	to the time	187.5′	242.5′
60	66′	you apply	270.0′	336.0′
70	77′	the brake	367.5′	444.5′
80	88′		480.0′	568.0′

WHEN DO TWO PLUS TWO NOT EQUAL FOUR?

In chemistry, 2 plus 2 do not always equal 4. That is because of the general nature of some substances and the peculiar way their molecules have of combining with one another. In some substances the molecules are separated to a greater degree than they are in other substances, and the mixing of the two different substances can give a surprising result.

For example, if you dissolve 6 cubic inches of sugar in 10 cubic inches of water, you will not get 16 cubic inches of sugar water. The result will only be 13.5 cubic inches. Nor will you get 4 cubic inches of dilute alcohol by mixing 2 cubic inches of water with 2 cubic inches of alcohol. The reason is that the sugar molecules, when dissolved in the water, slip in between the water molecules to reduce the whole volume to some degree. The same is true of the alcohol and water molecules. The final volume will not be the anticipated mathematical result you expected.

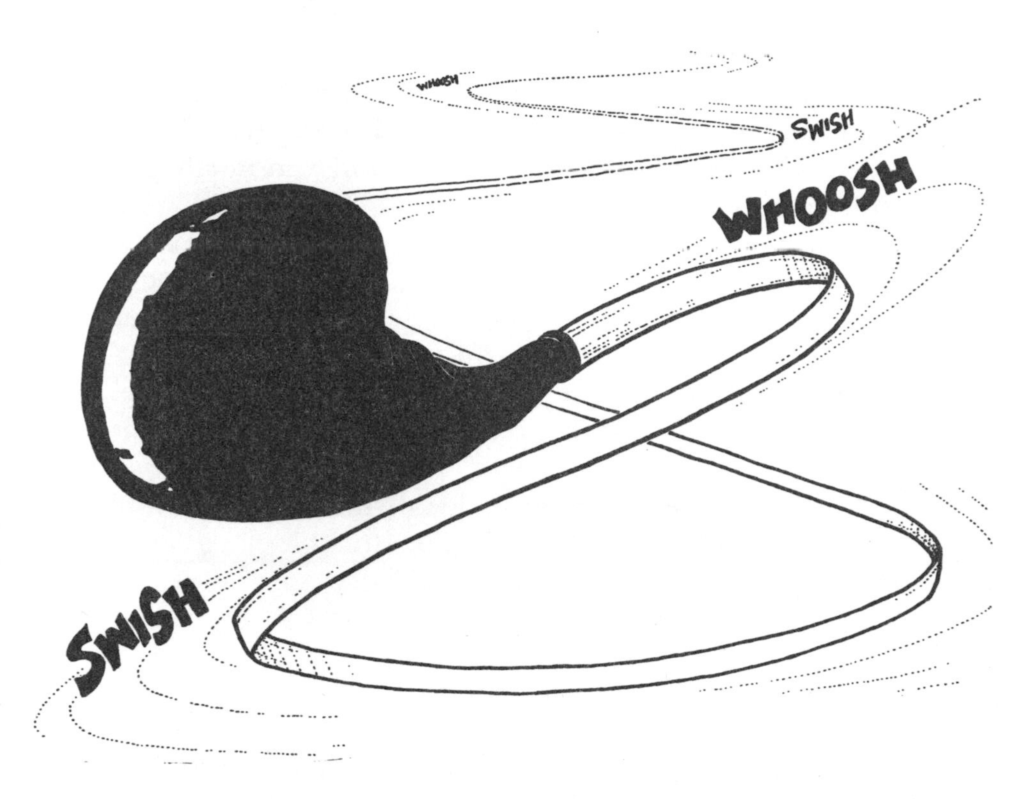

HOW DO THE JETS FLY WITHOUT PROPELLERS?

It is one of Newton's laws that says "For every action there is an equal but opposite reaction." Thus, the action of the air escaping from the mouthpiece of a toy balloon will cause an equal but opposite reaction which will make the balloon fly forward. It is this very same principle that makes the jet plane fly forward. It's the escaping gases from the rear of the jet engines that push the plane forward.

WHAT ARE LONGITUDE AND LATITUDE?

If the ship you were on was in trouble, rescue could reach you only if your captain could tell what his position was. He would have to state whether he was north or south of the equator. To do this he would use the imaginary lines called parallels of latitude, which run around the earth parallel to the equator, and north and south of it.

From the equator to the north or south pole would be 90 degrees, because either way it would be reckoned as only a quarter circle. To determine latitude at sea, the captain would use an instrument called a sextant which measures the angle in degrees between the height of the sun and its position above the horizon at noon.

To find longitude at sea, a local solar-time clock is compared with a Greenwich solar-time clock called a chronometer. One hour difference between the clocks is equal to 15 degrees longitude.

HOW DO WE TASTE FOOD?

When Fred's mother sat down at the table to relax and share the meal she worked so hard to prepare, ten-year old Fred angered her by saying, "This chop don't taste nothing." It wasn't the grammar but the ingratitude that made her angry. But Fred was telling the truth, because his head cold was so bad, it made his taste buds useless. It is well known that the senses of sight, smell and feel all enhance the sense of taste—especially the sense of smell. When saliva dissolves the molecules in the food, the taste buds which line the tongue's papillae receive the flavor. Two large nerves underneath the tongue then deliver the taste sensations to the brain by way of nerve cells. But since smell has so great an influence over taste, the message to Fred's brain was a limited one, due to his head cold.

The four primary tastes, from which all other taste blends are derived, are: sweet and salt (at the tip of the tongue); sour (at the sides of the tongue); and bitter (at the back of the tongue).

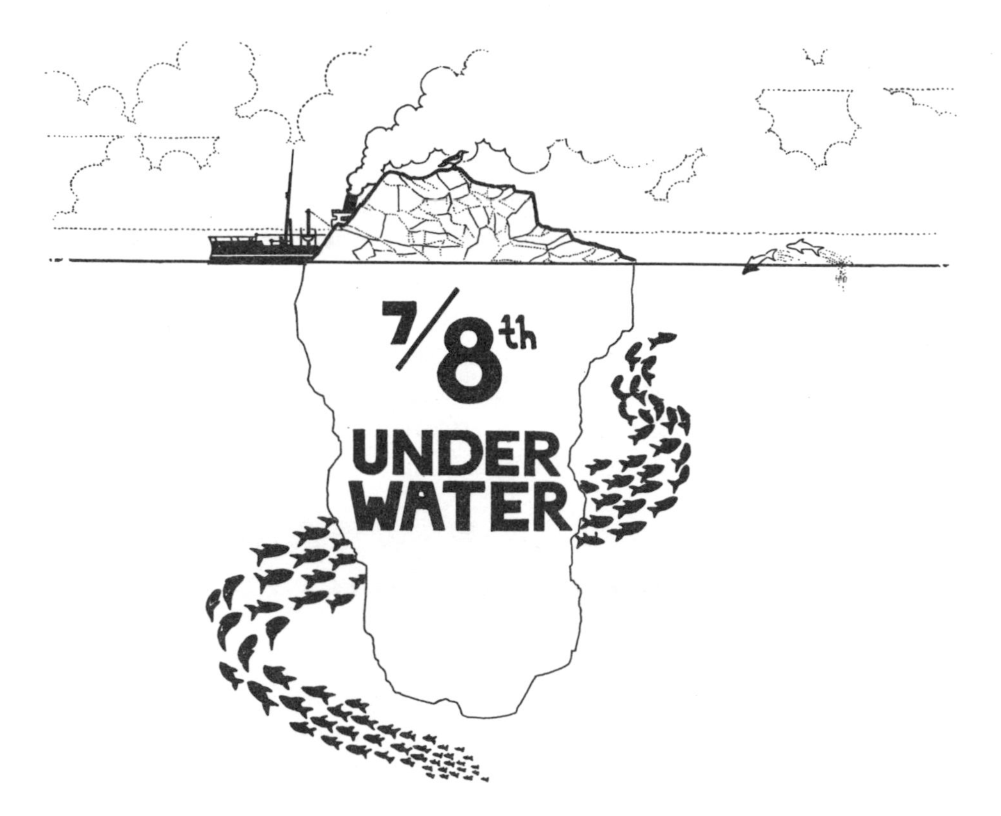

WHY DOESN'T ICE SINK INSTEAD OF FLOAT?

It would be just too bad if the floating phenomenon of ice were in reverse. Let us take a lake for example. If, in turning into ice from water, the ice were to become heavier, than the top layer of the lake would sink to the bottom. The exposed layer of water would then do the same, and soon the entire lake or river would be a solid mass of ice. All the fish would lack mobility and the surrounding terrain would become too cold for living creatures.

It is also doubtful whether in a short summer period the sun's rays could melt the entire lake before winter set in again.

Water increases its volume when it freezes, by one-eleventh, so that 11 cubic feet of water become 12 cubic feet of ice. Most substances contract when cooled, but when water is cooled below 4 degrees Celsius it expands. The expansion makes the ice lighter and that is why it floats. A cubic foot of water weighs 62.5 pounds while a cubic foot of ice weighs only 56.9 pounds—a difference of about 5$^{1}/_{2}$ pounds. The change of water into ice creates a force so powerful that water pipes break, car radiators crack, and soda pop and milk bottles explode. Also, great chunks of rock come tumbling down mountains, after splitting off by the expansion of freezing water between the cracks.

WHEN DOES THE FIRST HEARTBEAT OCCUR?

At the age of three weeks, a tiny bit of humanity, a thing we call a human embryo, has a heart in a beginning stage. At this time it looks like a single tube. Yet, though only three weeks old, it begins to beat. This starts an actual blood circulation through the first few blood vessels already formed around it. No matter how you try to explain the first heartbeat, it will always seem like a miracle.

WHAT IS EPILEPSY?

Epilepsy is a disease of the nervous system whereby the victim has a seizure and often becomes unconscious. It is not really an inherited disease, but a person may have a predilection for the illness. There are two kinds of epilepsy, petit mal and grand mal. In petit mal cases, the patient has brief lapses of consciousness and loss of control of parts of the body. This happens very frequently, even as often as once a minute. In the grand mal cases, the patient has a total loss of consciousness and rocks with convulsions for long periods of time. These attacks do not occur often, and may occur only once or twice a year. No one yet knows what causes the brain cells to bring on a seizure. Scientists agree that it could be from a brain injury before, during, or after birth. The patients are usually treated by giving them drugs which prevent and control the attacks.

HOW DOES A SODA-ACID EXTINGUISHER PUT OUT FIRES?

This type of fire extinguisher does not put out an electrical or oil fire. In either of these two types of fire, the soda-acid extinguisher (so called because the mixture creates carbon dioxide gas) will only spread instead of stifle the fire. In other types of fire, when the extinguisher is turned upside down the acid, in a small, loosely covered container, reacts with the bicarbonate of soda to create carbon dioxide gas. The mixture creates a gas pressure which drives out a stream of water mixed with the gas. The fire goes out for two reasons. The water cools the burning substance to a point below its kindling temperature, and the gas, which does not burn and is 1½ times heavier than air, suffocates the fire by settling down and driving out the oxygen.

WHAT IS A BRAIN WAVE?

Experiments have shown that the brain gives off waves of electrical impulses at regular intervals. They can be measured by attaching electrical connections to the surface of the skull. Measuring brain waves is called electroencephalography.

Brain wave patterns vary among different people and in different activities. Alpha rays are given off at a rate of ten per second when the brain is at rest. Beta rays are given off at a rate of twenty-five per second when the brain is active.

Brain waves of people vary with their personalities. Members of the same family often have brain waves that resemble each other.

Epilepsy and brain tumors give off certain patterns of brain waves. Sometimes these diseases can be diagnosed by means of brain waves. But brain waves can be used to detect only diseases that affect the cortex of the brain. Even the excitement of learning about a brain wave is enough to give your brain a little extra wave. But it will be a beta wave because that's the kind that's given off when the brain is active.

WHAT IS A SHOOTING STAR?

What we call a shooting star is really a meteor traveling through space at tremendous speed. Basically it consists of nickel, iron or minerals, and may have been a part of a planet which exploded in our solar system. This particle, when it comes close enough to the earth, is pulled toward it by gravity. In falling into our atmosphere, it catches fire in flight due to the friction in our oxygen-rich atmosphere. Most of these shooting stars or meteors burn up in the atmosphere. Few ever reach the earth. But when they do, what is left of them is called a meteorite.

As recently as October 27, 1973, a three-pound meteorite struck through the roof of a garage in Canon City, Colorado, with a speed of about 500 miles per hour, and shattered into 47 pieces. Glenn Huss, director of the American Meteorite Laboratory, said it was the most highly crystallized meteorite ever seen. Part of an asteroid swarm from a planet disintegrated eons ago, it was composed primarily of magnesium, olivine, iron, pure pyroxene, and crystallized troilite.

WHY DO WE MIX METALS?

Whenever two metals or more are mixed together chemically we have what is called an alloy. Iron by itself is a comparatively soft metal and cannot, for example, be used in the construction of buildings like the alloy known as steel. Steel is an alloy because it is made with iron and an addition of manganese and phosphorus while in its molten state. The resultant alloy of steel is so strong that it can be used to make the framework of tall skyscrapers. But alloys are made for many other reasons. Gold is too soft for handling in its pure state, so 10% copper is added to harden it. Other alloys are made to give the metal beauty, durability, ductility, malleability, hardness, rust resistance, resistance to strain, and so forth.

One student suggested an alloy of pancake flour and popcorn, so that the popcorn would flip the pancakes over by themselves.

HOW DO WE HEAR?

Our hearing apparatus consists of three main divisions: the outer ear, the middle ear and the inner ear. The outer ear is the least important and is there only to catch the sound. At the end of the canal is the eardrum, a membrane that vibrates when the sound waves strike it.

Behind the eardrum is the middle ear, a small hollow filled with air and three small bones connected to each other by ligaments. The three small bones in the order in which they are attached to one another are the hammer, the anvil and the stirrup. The stirrup is held to the thin rear wall of the middle ear. When a sound wave strikes the eardrum, the hammer, anvil and stirrup bones are set in motion. The motion continues on to the middle ear and into the walls of the cochlea. This is a snail-like compartment filled with liquid and the ends of hearing nerves. The nerve endings send their impulses through the auditory nerve and off to the front part of the cerebrum, which is the hearing center of the brain. The message is received and sounds are recognized.

IN WHAT WAYS IS THE CAMERA LIKE THE HUMAN EYE?

A science teacher likes to compare the camera to the human eye because they both function under the same principle.

When you close the shutter of the camera to keep out the light, it corresponds to the eyelid performing the same function.

When you adjust the diaphragm of the camera to admit the degree of light you desire for the best picture, your eye, through the automatic function of the iris, does the exact same thing. It is controlling the degree of light which is most comfortable for your vision at that particular time and place.

Of course the lense of the camera tries its best to emulate the human lens, but it can only go just so far. To focus the camera correctly, the person holding it must go back and forth till the distance is correct. But the human lens has ligaments and muscles to expand and contract it automatically, in order to get the perfect focus.

The film of the camera upon which the image falls corresponds to the retina in the back of the eye. It is upon the retina, a delicate tissue, that the image of the object we see falls. This is where the similarity ends.

The camera may be like the eye, but it will never be able to flash back all the beautiful pictures we wish to rerun in our mind's eye.

WHAT IS AN ALBINO?

An albino is a person or an animal whose skin, hair and eyes have no pigment or dark coloring matter. The melanocytes—cells located in an outer layer of skin—make the pigment that gives the skin its color.

Albinos may come in varying degrees. A complete albino has milky-white skin and hair, and pink eyeballs. The normal iris of the eye contains a pigment which gives it color. But since there is no pigment in the eyes of albinos, the eyes appear pink because the blood of the capillaries shows through the transparent parts of the eye. In the normal eye, light can be partially excluded by the pigment in the iris. But without this pigment to help exclude the light, the albino must squint or wear sunglasses.

This lack of ability to form pigment in the body is called albinism, and is an inherited condition. Albinism is a recessive factor and can show up in any generation. There is nothing in the records to show that albinos are physically weaker than non-albinos.

Albinos have appeared in all races all over the world including the Indians and the Negroes. Among the animals there are hereditary groups like the white mice, rabbits and poultry. Even flowers among certain plants have appeared as albinos.

WHERE DID THE PLANETS COME FROM?

No one knows the true answer but many wise men have proposed certain hypotheses or theories about their origin.

THE NEBULAR HYPOTHESIS. A French astronomer by the name of Laplace believed that a mass of hot gas or nebula started to cool and rotate speedily. While thus rotating it flattened out and formed nine rings around it which finally broke away to form the planets. The smaller fragments formed moons, comets and planetoids. It was the spiral nebulae in space that gave birth to this theory.

THE PLANETESIMAL HYPOTHESIS. This theory, which followed the NEBULAR theory, was advanced by two astronomers from the University of Chicago, Thomas Chamberlin and Forest Ray Moulton. It was their belief that a very large star, greater than the sun itself, passed very close to the sun and caused a huge mass of gas to rise by gravity. This was billions of years ago. After the star passed the sun and left on its way, the mass of gas cooled into small planets. By gravitational force, many joined together into larger bodies. These formed into the planets we know today, still in the gravitational pull of the sun.

THE PLANETESIMAL HYPOTHESIS (with slight change). Many astronomers believe in the passing-star theory but do not agree that the planets increased in size by capturing planetesimals. They believe that the planets were almost their present size when pulled off from the sun. They believe that the satellites were torn out of the planets by the pull of the sun.

CAN WE PREDICT EARTHQUAKES?

Scientists so far have not been able to predict earthquakes. The suddenness with which they appear makes them as difficult to predict as tornadoes. They come in waves that move at speeds of 3.3 miles per second in surface rock, and 4.8 miles per second in the next layer of rock. An instrument called a seismograph measures the severity of the quake. But there is no instrument to date that can predict an earthquake.

American and Russian scientists are working on a theory which could possibly predict an earthquake, but only in regions where small earth tremors precede large earthquakes. Unfortunately, large earthquakes occur without any advance warnings, and that is where the predicting difficulties lie. California has many different earthquake detection devices, but Professor Clarence Allen of the California Institute of Technology thinks that we'll never learn to predict earthquakes unless we can trap a few. To do this he says we'll have to blanket California with many more than the 500 instruments now in use. China, a country that suffers frequently from the most devastating earthquakes, has enlisted all its farmers and citizens as part-time observers to help make predictions. They have a theory that animals behave restlessly and nervously prior to an earthquake. They also watch deep water wells for a murky discoloration, believing it to be a sign of an impending earthquake. Thus far they have made some successful predictions and also many false alarms.

WHAT IS THE WATER CYCLE?

The water cycle is a name given to the constancy of the water content of the earth. When water rises into the atmosphere as a vapor or fog, from the oceans, seas, rivers and lakes, it always returns to earth again as a liquid in the form of rain, sleet, snow or hail. This constant change is called the water cycle.

WHAT IS THE NITROGEN CYCLE?

Nitrogen gas, which makes up about 78% of our atmosphere, is constantly being used by plants and animals. It is then cast off into the atmosphere, only to be used and cast off again. That is why it is known as the nitrogen cycle. Soil bacteria change the nitrogen from the air into nitrogen compounds in the soil. This is used by plants to form proteins. Both man and animals use these proteins by eating the plants. When man, animals and plants die and decay, the nitrogen goes back to the soil and then into the air again. This is called the nitrogen cycle.

WHY DOES PENICILLIN MAKE SOME PEOPLE SICK?

No two people are built alike either in physical appearance or in internal chemical makeup. The various chemical differences in the bodies of people are the reason for the different reactions to penicillin. Edema is a swelling due to excessive tissue fluids. If a person cannot take penicillin, an allergic reaction takes place because the penicillin molecule causes a leakage of serum from the blood vessels, causing pulmonary edema and even death.

IS THERE LIFE ON OTHER PLANETS?

Having explored the moon and found no life on it, science is now trying to explore planets in outer space to detect any life on them. We knew beforehand that life could not exist there as we know it here on earth, because the planets lack water and air. In our form of life, these two are basic essentials to plants and animals. Without oxygen or water, life cannot exist on this planet.

There are certain figurations on Mars which were thought to be caused by living things. But we are not convinced that life such as we know it is possible on Mars. We believe that this earth is the only planet where life exists.

Every now and then, some unknown foreign object is claimed to have entered our planet from outer space, and to have been seen by someone. Yet there is no definite proof that any life was on it, nor that it came from a planet where life exists. But it does seem unlikely that life should have arisen and flourished on only one tiny planet like the earth, when the universe is so incomprehensibly vast. As of today, we haven't even an inkling of an answer.

Perhaps the present probes that are now being developed by NASA scientists, to get evidence for bacterial life on Mars, may someday convince us that there is life on Mars.

WHAT ARE THE NORTHERN LIGHTS?

Northern lights are also known as Aurora Borealis in the Northern Hemisphere and Aurora Australis in the Southern Hemisphere. It is a well-known but little-understood phenomenon—a glowing or flickering of light of natural origin sometimes seen at night in the sky. It is too faint to be seen except in the far northern or southern regions. The beams of light, which are best seen at night, keep changing positions and brilliance from second to second. At full intensity, the light covers the entire sky from about 50 to 600 miles up.

It is believed that the lights are caused by rays of electrically charged particles, such as electrons, shot from the sun at high velocities, which are affected by the earth's magnetism and may collide with gases in the earth's atmosphere.

The light appears within a few hours after a magnetic storm on earth. The aurora is especially common during years of sunspot activity and is rare in other years.

IS ASTROLOGY A SCIENCE?

If we apply a definition of science to astrology, we must conclude it is not a science. Scientific knowledge must be exact and the results the same anywhere under the same conditions. Conclusions are arrived at only after many testings, investigations and experiments. Astrology does not fulfil these conditions although its origin was noble and goes back as far as written history can record. It even gave birth to the real science of astronomy.

Ancient Babylonians, and later the Egyptians, Romans and Greeks, first studied the motions of celestial bodies. It was when they started to relate celestial studies to human behavior and destiny that they lost their scientific aspect. During the fourteenth and fifteenth centuries astrologers were important in the courts of Europe. Shakespeare called them soothsayers in Julius Caesar, but even as late as the seventeenth century they were consulted on the diagnosis of the sick, prediction of lucky hours, and the best site to build a castle or a church or other municipal buildings.

The horoscope of a person is a map of the position of the planets, the sun and the moon at the moment of his birth. The signs of the twelve divisions of the zodiac into which the planets appear are supposed to indicate the influences on the future of that person. But how scientific can this be, how exact the predictions, when a study of thousands of twins, who obviously had the same horoscope, shows that their careers and destinies were miles apart? How scientific can it be when the destiny of people is foretold by people who cannot even tell what tomorrow holds for them? That is why, in his Confessions, St. Augustine said that he gave up his belief in astrology when he learned that a wealthy landowner and a slave on the same estate were born at exactly the same time.

IS A COMET A LARGE METEOR?

The answer is no, for several reasons. In the first place, a meteor is a fragment that broke away from some celestial body and is heading towards our earth. The comets do not come towards the earth to burn up or fall into it like meteors do, but fly around the sun at tremendous speeds. A meteor may be very large, about a mile wide in diameter like the one that formed the crater in Flagstaff, Arizona, or it can be the size of a grain of sand. But a comet has a head, and its tail can be hundreds of thousands of miles long. The meteor is a solid of either iron, nickel or stone, but the comet consists of pieces of solid matter mixed with masses of frozen snow and ice crystals. A meteor has no orbit to travel in, since it is a falling mass, but a comet travels around the sun it its own orbit. A meteor has no head or tail, but a comet has a head. It also assumes a tail as it approaches the sun. This tail always points away from the sun as it goes around it.

A most exciting example is Halley's Comet, named after its discoverer. This comet, last seen in 1910, appears about every 76 years.

WHAT CAUSES A HURRICANE?

Although hurricanes are the most devastating storms that rake the earth, with winds that can reach 230 miles per hour, how they are spawned is still a mystery. Weather scientists still aren't sure what chain of events must be linked together to change a minor weather disturbance into a roaring hurricane. One scientist defines it as "a huge atmospheric heat engine, sucking in air at its bottom for fuel, and spewing out air at its top like a car spews spent fuel from its exhaust pipe."

Hurricanes are virtually nonexistent in the North Atlantic because it is a cold-water region, and a hurricane needs tropical oceans to pull in the moist warm air from the ocean surface towards the low-pressure center called the eye. The eye is the calm portion of the spiral, and remains so throughout the storm.

Many disturbances die out, yet one of several hundred slowly assumes the tremendous spiraling motion that turns it into a full-blown hurricane. Why so many fade out and a few do not, we still don't know. As the storm grows it takes on a forward movement that can range from a few yards per hour to 60 miles per hour, and in its mature state can dominate thousands of square miles and sustain itself for weeks. The damage it can do to cities runs into the millions, and, roughly speaking, the amount of energy it can release in one day could supply electricity for the entire nation for about six months.

WHAT CAUSES OUR EARTH TO QUAKE?

Our cool earth was once quite hot. In fact, there is plenty of boiling and sizzling still going on, way down in the innards of the earth. In the cooling-off process, which is still taking place, the surface is cracking and wrinkling much like the skin of an apple drying up with age. Volcanoes and earthquakes are connected with the same set of causes. The earth often trembles because of sudden movements in the rocks below the earth's surface. Many different forces affect the inside of the earth. These forces push against rocks till they can resist no longer. Then they give or stretch or break along lines called faults, and an earthquake results.

Sections of rock slip past each other and set up vibrations which cause the earth to tremble. An apparatus that registers the shocks and motions of earthquakes is called a seismograph. Any tremor registering 4.5 or more on the Richter scale is considered potentially dangerous.

Sometimes the tremendous underground forces push up mountains, and sometimes this occurs in reverse, when the mountains settle, the earth may slip and cause an earthquake. But no matter what the cause, earthquakes are one of man's greatest fears.

WHAT ARE SOME IMPORTANT OCEANIC DISCOVERIES?

1. All the oceans except the North Pacific are divided in the center by an almost continuous system of mountains.

2. Fish thought to be extinct for 50 to 70 million years were discovered thriving off South Africa in 1938.

3. A layer of living organisms lies on the ocean floor several hundred fathoms down.

4. Nodules of manganese, cobalt, iron and nickel can be dredged from the floor of the seas.

5. The ocean bed is underlaid with basalt, not granite.

6. There are deep sound channels that can carry sound for thousands of miles.

7. There is life in the deepest parts of the ocean.

8. Man can live and work in the ocean for extended periods of time. This led to the concept of underwater habitation by Jacques Cousteau and Edwin A. Link.

WHY DOES A SALTY OCEAN GIVE SWEET RAINWATER?

When salt water evaporates, only the water molecules leave the ocean to enter the air. The salt, being much heavier, remains in the ocean and does not rise in the vapor above the lake or ocean.

WHAT GIVES THE OCEAN ITS DIFFERENT COLORS?

The colors of the ocean are due to materials that are dissolved in the water, and to the different colored microscopic living things that inhabit the waters.

The waters may be red because microscopic plants called dinoflaggellates are there at the time. These plants can poison the water for many kinds of marine life during a "red tide."

The waters may be brown due to brown microscopic algae that form weeds two hundred feet long. The brown algae known as "devil's apron" are food for whales.

The waters may be green due to green algae which are abundant in the oceans and which are a source of food for fish.

Another mat-forming alga that floats and gives color to the oceans is known as gulfweed and is greenish in color.

Last but not least is another ocean-floating mass of algae known as plankton.

All these things in different places at different times give our vast oceans their different colors. But wherever the water is free of pollutants, and wherever the light can penetrate to the greatest depths, the water will always appear blue.

WHAT IS LIGHTNING AND THUNDER?

When your static-loaded finger touches the refrigerator handle, you create a spark. That's how lightning occurs. The friction of air on raindrops gives them an electrical charge, just like the rug friction gives your body a charge. The raindrops break into smaller drops separating the balanced into an unbalanced electrical charge. The air that separates the cloud from the earth acts as an insulator and resists the efforts of the opposite charges of electricity to rush together. When the strain becomes too great, and there is a sufficient charge, one equal to billions of volts, a lightning bolt leaps from one cloud to another, or from a cloud to the ground. A bolt between clouds may be as much as twenty miles, and a bolt between clouds and the earth could be over a mile long.

The lightning bolt forces the air apart and creates tremendous heat in the air in its path. The sudden heating causes the surrounding air to expand violently. When the air rushes back into this partial vacuum created by the bolt, the resulting noise is the thunder or the great air wave caused by the inrushing air.

If you watch the lightning streak and count the seconds till thunder noise reaches you, you can tell the distance the storm is from you. Thunder travels approximately one mile in five seconds. Therefore, if it takes fifteen seconds, you know that the storm is three miles away.

WHO BUILT THE WHITE CLIFFS OF DOVER?

The white chalk cliffs of Dover in England were built by billions of colonies of tiny protozoan marine animals known by the name of foraminifer. They left these white chalk deposits long after they died. These "foram shells" are composed of calcium carbonate ($CaCO_3$) and at one time were used to make blackboard chalk. Forams lived in extremely large numbers in the primitive ocean. After their death, their remains went to the bottom of the ocean floor. Millions of years later, these mountains rose out of the water, giving the world what is now known as the White Cliffs of Dover.

WHAT BRINGS ON THE HAILSTONES?

Hailstones are lumps of ice that fall from the clouds during a thunderstorm. They have a center of snow around which are alternate layers of clear and snowy ice. They vary in size from a pea to a walnut, and take many shapes in their formation.

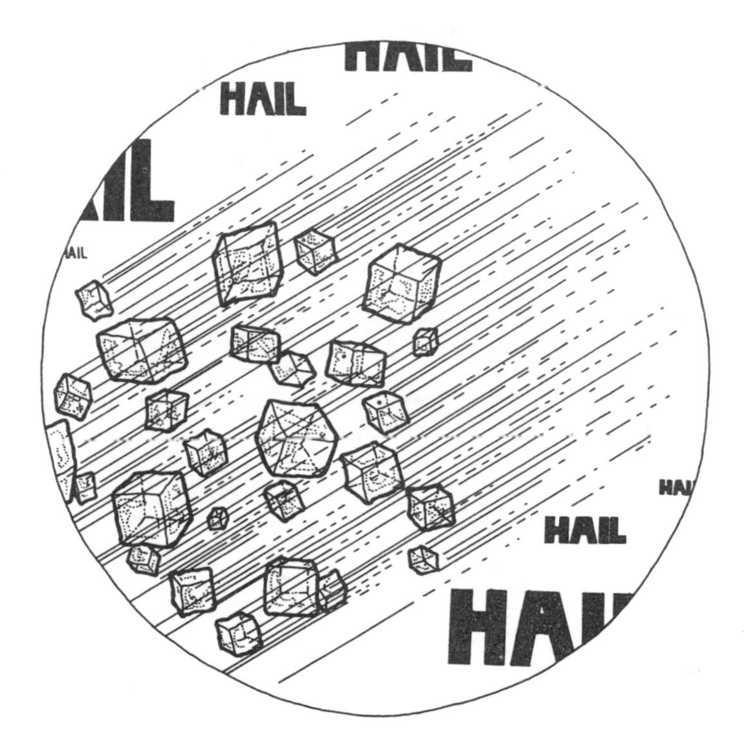

Hailstones usually occur in warm weather and last a short while. They occupy a narrow path within a thunderstorm. When the air on the earth's surface is overheated, and the air above is cold, wind currents develop that move rapidly up and down in a vertical motion. When the warm air containing moisture hits the cold, below-freezing temperature above, ice crystals form. As these crystals are blown up and down, layer upon layer of crystals form on the hailstones, until they weigh so much that they fall to earth because their weight keeps them from staying in the air any longer.

That's the time for you to get out of the way.

THROUGH WHAT ELEMENT DOES SOUND TRAVEL FASTEST?

Ordinarily one would guess that since air is rarer or less dense than water, wood or steel, sound would travel fastest through it. But just the opposite is true. Sound travels through air the slowest of all. Through air it travels 1,129 feet per second, through water about 4,760 feet per second, through glass and steel about 16,000 feet per second. Since the air molecules are not as closely packed as water, glass or iron molecules, the sound waves lose a great deal of their energy while passing through it.

HOW DO GLACIERS FORM?

Many glaciers that remain upon our earth today are left over from the recent Ice Age which lasted until about 25,000 years ago. The coldness of high altitudes causes mountain glaciers to melt more slowly than glaciers on lowlands. There still remain about five million cubic miles of ice around the North and South Poles. No one really knows how glaciers are formed, but there are several theories about it. One theory is that since volcanic activity brings about colder winters and shorter summers, there may be a connection between the two.

Glaciers are huge sheets of ice, hundreds of feet thick, that travel slowly over great distances, changing the earth's physical features as they travel. If they break away in large chunks and land in water, they become icebergs. But as they move slowly over land, they pick up and carry rock and soil with them, thus eroding the land. If they reach an area of high temperature, the glacier melts and deposits its load of rock, gravel and soil over a wide area. When they pass through river valleys, they deepen them. They also wear down and round off tall mountains. Many lakes were formed by the large holes dug by glaciers, which later were filled with water.

All the pressure of the overburdened mass of ice causes the bottom to liquefy and provide a film slippery enough for the ice to move downhill. This process is called regelation.

HOW MUCH ENERGY IS THERE IN A WAVE?

The power that is carried by a wave is important information to a structural engineer. We know that the kinetic energy is tremendous. A four-foot, ten-second wave striking a coast expends more than 35,000 horsepower per mile of coast. On the coast of Scotland, a block of cemented stone weighing 1,350 tons was broken and moved by the waves. A replacement weighing 2,600 tons was also carried away. Finally the engineers concluded that the force of breakers along the coast was 6,000 pounds per square foot.

HOW DOES THE ELECTRIC-EYE DOOR WORK?

To children who go through these doors it may seem like magic. But this magic takes place because scientists discovered that several elements like potassium, sodium, selenium and lithium give off electrons when rays of light strike them. Electrons are negatively charged electrical particles. If a positively charged substance can be placed near the negative electrons that come off the element hit by the light, you will have an electric current when the two opposite particles meet. Of course the current will not be strong.

All this is accomplished be a globelike device

known as a photoelectric cell. This globe or bulb has half of its inside coated with the material that gives off electrons when light hits it. Inside the tube is a positive-charged plate (a curled copper wire) fixed in such a way as to catch or attract the negative electrons as they bounce off the coated side of the tube.

The small current that is created can be amplified to operate a strong motor to open a door. Thus, when a light beam is directed across a doorway, it is picked up on the other side by the electric eye and run to a motor that keeps the door closed. But a person crossing the path and breaking the beam with his body releases the motor from its duty of holding the door closed, and thus permits the door to open.

HOW STRONG IS A PERMANENT MAGNET?

A permanent magnet made of an alloy (mixture) of cobalt, nickel, and aluminum can hold up to 60 times its own weight. A 60-pound boy, suspended in steel straps, can be lifted off the ground by this one-pound magnet. The alloy is called alnico.

CAN MAN'S LIFE SPAN BE EXTENDED?

During the time of the American Revolution, the average life span of man in the United States was 39 years. Of course, this doesn't mean that there weren't people who lived to 80, 90, or even 100 years. It just means that disease, and the high rate of death in childbirth, and man's inability to cope with it from a scientific standpoint, prevented the average person from living beyond 39 years.

Recent experiments with cell tissue by Dr. Leonard Hayflick, a Stanford University microbiologist, indicate that the death of human cells is "programmed" into the cell. After 11 years of frozen storage, the cells maintained their "memory" of where their division left off.

Throughout history, man's life span of 80 to 90 years has remained about the same. Scientific discoveries and advancement just made it possible for more people to live out their life span. But in reality, it has not been extended.

The latest statistics from the 1979 Almanac show that males born in the United States in 1976 have a life expectancy of 69.7 years, and females 77.3 years.

WHAT DECIDES THE SEX OF THE UNBORN BABY?

Complete studies of the body cells of humans, particularly as regards the types of chromosomes in the cells, has led to the conclusion that it is the male cell that decides whether the newly fertilized ovum or egg will produce a male or female child.

All humans have either two X chromosomes in their body cells, in which case they are female, or an X and a Y chromosome when they are male. These X and Y chromosomes are known as the sex chromosomes.

The female human egg always has one X chromosome when unfertilized. But when fertilization takes place, upon the entry into the egg of a live male sperm, it is at that precise instant that the sex of the future child is determined.

If the sperm carries an X chromosome, this joins the X chromosome of the ovum or egg, to give the unborn fetus a female sex, because the fertilized egg now contains two X chromosomes. If the sperm had contained a Y chromosome and joined the X chromosome of the egg, then the XY would make the fetus a male.

DOES THE TELEPHONE WIRE CARRY YOUR VOICE?

The mouthpiece or transmitter and the receiver to the ear have a small box filled with tiny carbon granules, and a thin metal disc called a diaphragm. Sound waves vibrate the diaphragm which alternately presses the carbon grains together and then loosens them. There is an electric current in the telephone wires between your telephone and the other party.

When the diaphragm presses the carbon grains together, more electric current flows in the circuit. When the grains are farther apart, there is less current. In the receiver there is another diaphragm, which rests on a magnet. The strength with which the magnet pulls on the diaphragm varies in proportion to the amount of the electric current. The back-and-forth movement of the diaphragm generates waves that reach the ear as sound. Because the vibrations in the diaphragm in transmitting control the flow of the current, they produce identical vibrations of the diaphragm in the receiver. Thus, sound waves that leave the receiver are identical with those that enter the transmitter.

Now the mystery is gone. The wire does not carry voices but electrical impulses which are later transposed into sound.

HOW DID THE COLORED SEAS GET THEIR NAMES?

Many children at school wonder why this sea was the Red Sea and that sea was the White Sea, but they hesitate to ask the teacher why. Here are the answers.

Whenever floods occur, yellow mud is carried into the sea, giving it a yellow color. That is how the *Yellow Sea* got its name.

The *Black Sea* has no outlet, and because it is entirely landlocked, its deficiency in oxygen at a depth of 200 meters gives it a high concentration of hydrogen sulfide. This comes from the decomposed bacteria that drift down from above, resulting in a black color. And so the Black Sea got its name.

The *Red Sea* got its name because there is an ever-recurring bloom of small algae that gives the sea its permanent look of red.

Because the *White Sea* is covered with ice for more than 200 days a year, and has a white appearance so much of the time, it is called the White Sea. Unfortunately, due to so much pollution in the Danube River, anyone taking a boat ride is lucky if he sees any blue at all in what is called the Blue Danube.

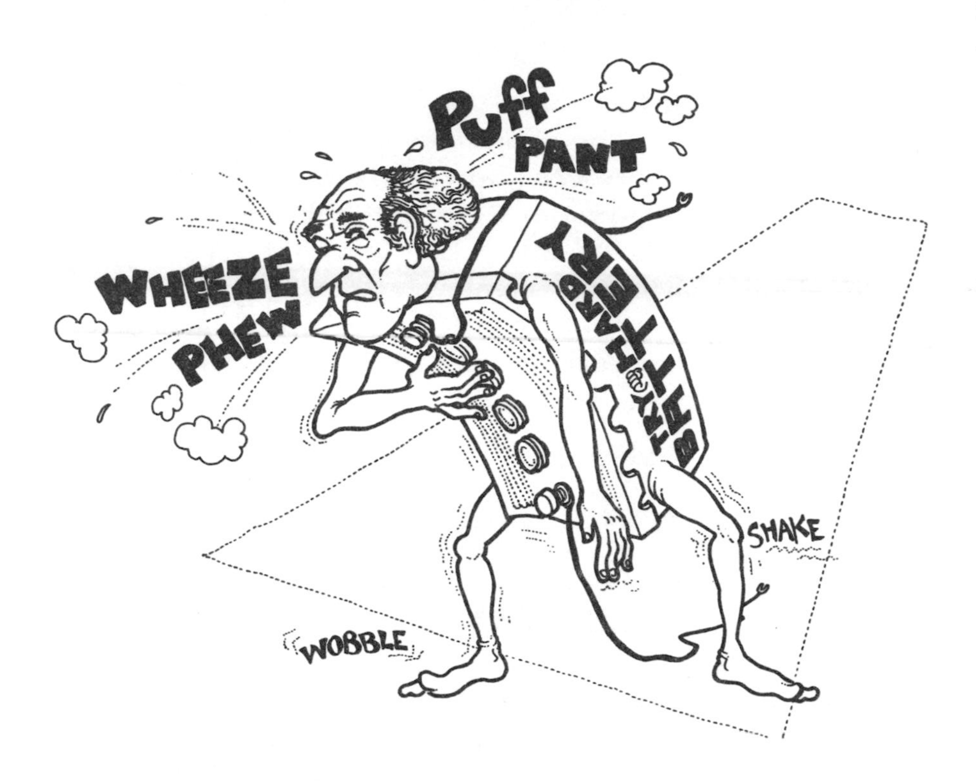

WHY DOES A CAR BATTERY GO DEAD?

A storage cell or battery creates a flow of electricity when lead plates, which have a negative charge, and lead dioxide–covered plates, which have a positive charge, are both suspended in sulfuric acid. It is called a storage battery because it stores electricity for future use. Each cell in the battery creates about two volts of electricity, and the greater the number of cells, the greater is the voltage.

A storage battery runs down when it is not used for a period of time or when it is overused by leaving the lights or the radio on in the car when the motor is not running and the generator is not recharging. If a generator in a car is not functioning properly, then even though the car is in use, the battery will run down because it is not receiving any current from the generator to recharge the used-up current in the battery. A rundown battery can be recharged by introducing an electric current into the battery for a few hours, thus giving it a fresh supply of negatively charged electrons. A battery will go completely dead, with no possibility of repair or recharge, and will deliver no current at all, when the plates or the chemicals are used up. At that moment it is good to remember that there is only one thing free of charge in this world, and that is a dead battery.

CAN FISH LIVE IN THE OCEAN DEPTHS?

Until 1860 no one believed that there was life in the oceans below 1,800 feet. But when a telegraph cable was retrieved from an ocean depth of 6,000 feet, a variety of marine life was found to cover it.

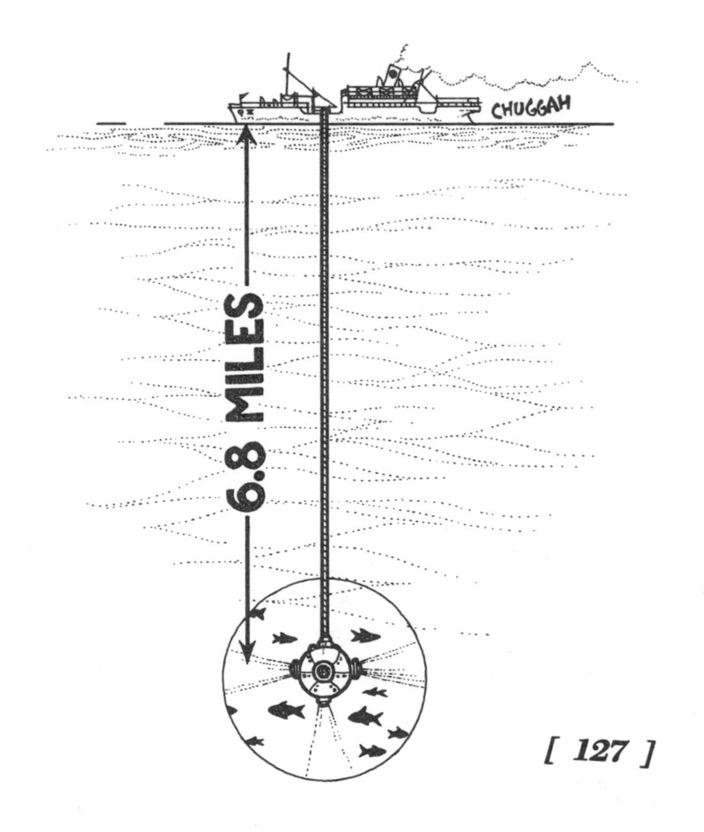

In 1951 a Danish oceanographic ship dredged various invertebrates from a depth of 33,433 feet in the Philippine Trench, and a year later caught fish at a depth of 23,400 feet.

In 1960 Jacques Piccard and a man named Walsh reported a flatfish resembling a sole at a depth of 35,800 feet, which is about 6.8 miles below the surface. The fish, observed from the porthole of their bathyscaph named Trieste, was about one foot long and about six inches wide.

WHAT IS ASBESTOS AND HOW IS IT IMPORTANT?

People often wonder whether it is a mineral or a rock. Well, they are right both ways, for it is a mineral and a rock. In nature it is found in the form of white fibrous rocks. The fibers, when separated, are blown loosely into walls for insulation, or pressed into paper. The fiber is also made into curtains and drapes. Since it is a very poor conductor of heat, it is made into shingles and boards and used in the building industry. The best brake linings for autos are made from this. To save heat, it is used to cover hot steam pipes, and to line hot blast furnaces. The chief source for this all-around mineral is the city of Quebec.

Unfortunately, it has lately been found that the tiny particles of asbestos which break off during commercial processing have caused a serious lung disease among the workers who were exposed and who inhaled them. And it was also found that the members of their respective families who breathed the particles that fell from their clothing were likewise inflicted with the lung disease. Now many workers are planning lawsuits to receive compensation for the disease they acquired while facing a hazard they were not told about.

CAN THE SEA BE MINED FOR GOLD OR DIAMONDS?

Gold and diamonds are presently being dredged from the ocean floors in several areas of the world. Since 1952 over half a million carats of diamonds have been dredged off the southwest coast of Africa. A United States company is dredging about $200,000 worth of diamonds a month from depths of 200 feet. They are found in mixtures of sand, gravel and boulders, but the operation is not yet profitable because dredging techniques to handle such large quantities of sand have not been perfected.

Prospectors are presently covering about a million and a half acres of the Alaskan coast in search of gold, and along the coasts of North Carolina and Oregon exploration is going on to see whether or not gold can be dredged economically in these areas.

WHAT PART OF YOUR BODY INFORMS YOU OF THIRST?

Eating too much salty food at one time raises the amount of salt in the bloodstream. Ordinarily the blood has a very fixed percentage of salt in it and does not welcome change. Our body has a concentration sensitivity device which signals the kidneys to absorb the excess salt in solution. By absorbing the excess salt, much water is taken away from the body tissues. Thus, one seeks more water to replace that which has been absorbed. The sense of thirst arises by stimulation of cells in the pharynx, due to the loss of water from the cells with which nerve endings are in contact. The decrease of concentration of water in the blood bathing these cells causes a drying of the lining of the pharynx because of decreased salivary secretion. Therefore the urge to drink.

IF WE CAN MAKE LIGHTNING, WHY CAN'T WE USE IT?

This is a good question and sounds logical. But to this day scientists have not learned to harness and control lightning, nor to store it for future use. They have learned how to make powerful lightning flashes, and how to ground electrical lightning flashes with lightning rods. But this is a long way from knowing how to trap it, store it, and use it when you want to.

HOW CAN YOU MAKE A MAGNET?

You magnetize a needle by stroking it on a magnet. The molecules in the needle, which were previously in a scattered arrangement, become arranged in a neat, orderly manner under the influence of the magnet and form north and south magnetic poles. Manufacturers of loud-speakers use more magnets today than any other industry.

WHAT IS THE CONTINENTAL SHELF?

The continental shelf is a sort of underwater ledge around the coastline of the oceans of the world. Long before electronic equipment such as sonar was invented, this shelf was thought to be flat. Now we know its contours and can draw an almost accurate picture of the shelf on all shores around the countries of the world. We know it has basins, ridges and canyons.

The edge of the shelf, where the bottom begins to slope steeply, is found at depths between 360 and 480 feet. Hills and basins on the shelf usually do not exceed 60 feet. The width of the shelf could vary from a few feet to several hundred miles. On the eastern coast of the United States it is many times wider (about 150 miles in the northern part) than that on the west coast (about 20 miles). It could be said that the average width of the continental shelf around the world is about 40 miles. The widest shelf is about 750 miles on the Bering Sea in the Pacific.

The slopes of the shelf drop gently an average of 12 feet per mile from the shore to the continental slope. But in contrast, the grade of continental slopes is 100 to 500 feet per mile. At the edge of the shelf there's a drop of two to five miles to the bottom of the ocean. At the present time, mineral explorations are going on in the areas of the continental shelf which underlie about 7% of the ocean. It is always hoped that new and better tidewater oil fields will be discovered under the shelf.

HOW DO JOINT BONES FUNCTION?

A joint is where two bones meet in your body in order to allow movement in various directions. The immovable joints of our body are the ones located in our skull. But even they will move slightly under great pressure to prevent too much damage. In our bodies there are four different kinds of movable joints.

PIVOT JOINTS. They help rotation and are located in your elbows and neck.

HINGE JOINTS. They allow a back-and-forth movement like a door, and are located in your elbows, knees, fingers, toes and neck.

BALL-AND-SOCKET JOINTS. They allow the greatest freedom of movement and are located in your shoulders and hips.

GLIDING JOINTS. They are located in your wrists.

Machine experts go to great lengths to see that movable parts are properly oiled to prevent friction. Have you ever thought about your body joints? Do you realize that in your lifetime you never even have to grease them once? Let Mother Nature explain. She made two provisions for the care and preservation of your joints. In the first place, she provided a thin layer of smooth, elastic cartilage to protect the bones from friction. To further enhance the smooth movement of the joints, she provided a liquid substance called synovial fluid to keep them moist and lubricated.

Now isn't our body a wonderful machine?

CAN HUMANS ADAPT TO RARER AND HIGHER ALTITUDES?

In Lake Titicaca, at the top of the world, lying in the Andes Mountains and equally divided by Peru and Bolivia, there lives a population of small sturdy Indians whose average weight is about 150 pounds. In this climate, so short of oxygen which is so vital to us, a newly arrived visitor gets headaches, feels slightly nauseated, and is immediately short of breath. Because of the lower oxygen supply, outboard motors generate only half their designated horsepower, and a cigarette lighter, which naturally needs oxygen to support its flame, lights with great difficulty. In fact, until recently forest fires were unknown there. But the strong-bodied Indians have physically adapted themselves to the scarcity of oxygen by developing lungs and chests far larger than normal, and carry a quart more of blood than the average human of equal size. Add also an extra million more oxygen-bearing red corpuscles, and you have the answer to how humans adapt to certain climates.

WHAT HAS AMBER CONTRIBUTED TO SCIENCE?

Amber is a fossil resin, the sap of ancient trees whose land became inundated and covered by the oceans millions of years ago. Later on, the sap was washed up on the shores of the Baltic Sea, in Denmark and in Sweden. The amber washed up in Sicily is brownish-red in color and fluorescent.

Since it was not affected by the chemicals in the ocean, amber has preserved 75 species of ants, bees, beetles, mites and moths that lived and crawled inside the trees millions of years ago. Trapped in the sap of the trees, these insects were carefully preserved for mankind to study today.

When the Greeks first found amber, and learned that it was able to acquire an electric charge when rubbed, they called the substance *elektron.* If a pipe has an amber stem, it is because amber is shapeless to begin with, is non-brittle in nature, and can be carved into jewelry, beads, trinkets and pipe stems.

COULD ACID IN FOOD BE HARMFUL TO US?

Yes, it could. If you ate too much of the same thing consistently, and in abnormal quantities—say, for instance, you went on a lime drink binge—a cumulative effect could result. But on the whole, the acids of most foods—and many of our foods contain acids—are helpful and useful. Our own digestive system contains hydrochloric acid to help break down certain foods. Here are some foods and the acids they contain:

Milk—Lactic acid
Apples—Malic acid
Fats—Oleic acid
Cider—Acetic acid
Butter—Butyric acid
Citrus fruit—Citric acid
Tea—Tannic acid
Grapes—Tartaric acid
Cranberries—Benzoic acid
Tomatoes—Oxalic acid

WHAT IS SO INTERESTING ABOUT THE GULF STREAM?

It is estimated that this unseen giant river that flows into and under the Atlantic Ocean, starting from the warm tropical waters of the Gulf of Mexico, is about one thousand times larger than the Mississippi River.

It is about 50 miles wide and about 2,000 feet deep and divides the Atlantic vertically and laterally, at a temperature of about 80 degrees. Its tremendous volume is a constant flow at about 70 miles per day, into the Florida Straits. Then it flows north to the Newfoundland Banks, where it ceases to be a current and drifts eastward to Europe. Here it divides, going to the west coast of Africa and north to the British Isles. This undersea river brings warmth and comfort to northwest Europe as well as to the southeast coast of Florida. However, in the Arctic region so thick a fog is created that navigation becomes hazardous and difficult.

WHY ARE SOME PEOPLE COLOR BLIND?

A person who is color blind lacks the ability to discern certain colors such as greens, reds and a few others. Color blindness is believed to be hereditary or sex-linked through the genes in the chromosomes. In most cases it is found in males, in a few instances among females.

The retina of the eye has two types of cells, called rods and cones. The eyes seem to take two pictures at one time when we see an object. The rod-shaped cells allow only variations of light to enter, but not color. They contain a substance called visual purple which helps the rods to see. The color picture is taken by the cone-shaped cells which are estimated to be about 7 million in number in the entire retina. Both rod and cone cells connect to the optic nerve. The optic nerve therefore receives a complete picture of the white and grays from the rods, and the color effects from the cone cells.

But the method of stimulation of the cones by light is not well understood. Pigments whose breakdown initiates nerve impulses have not been identified in the cones. But experimental evidence lends support to the idea that three substances may be present in different cones: one most sensitive to red light, one to green light, and one to blue-violet light.

The theory of color blindness therefore seems to be that the chemical color for red or green or whatever color one is incapable of distinguishing is lacking in the cone cells and therefore cannot be reflected.

HOW DO GENETICISTS EXPLAIN MONGOLISM (DOWN'S SYNDROME)?

Although mongoloid babies seem normal at birth, their abnormality soon appears. They lack nose bones and teeth, and usually have some heart disorder. They have flat noses on broad faces, with skin folds over their eyes. Most are seriously retarded and many later develop leukemia.

In 1956 the true nature of genetic disorders was first ascertained when scientists found a way to accurately observe human chromosomes. Then it became possible not only to count them, but to identify each of the 23 pairs contained in human cells.

During the gestation period, when cells are constantly dividing, the chromosome pairs are regularly splitting and separating in an orderly and precise manner. But sometimes one of the chromosomes misbehaves and violates this pattern. In 1959, three French scientists examined cells of children with Down's Syndrome and found that they all possessed 47 instead of 46 chromosomes. They concluded that in all the mongoloid cases, the condition was due to a breaking of the pattern of regular chromosome behavior during the process of cell division. What causes this irregularity in the chromosome pattern is still a mystery to geneticists.

HOW DO SOME LIVING THINGS GIVE THEIR AGE AWAY?

Trees have their annual rings, and a great number of fish have scales with rings that give their age away. The lines on the shells of clams, and in the horns of mountain sheep, also reveal their age. Intermittent growth of the horseshoe crab tell us its age, and the colorful bill-ridges of the puffin bird tell us how old it is. In its external auditory canal, a whale develops a waxy ear plug which shows annual lines of growth when sectioned lengthwise. In a sturgeon, the annual rings of growth can be counted in the strong cartilaginous rays that support the pectoral fins.

WHAT HAPPENS WHEN A DIVER GETS THE BENDS?

Bends is a painful condition caused by the formation of nitrogen bubbles in the bloodstream and body tissues. This happens when the air pressure surrounding the body is lowered too rapidly. When the diver comes out of the deep too quickly, nitrogen bubbles come out of the body fluids and enter the bloodstream and tissues, causing a tingling sensation at first, followed by pain in the joints, chest and abdomen. In bone and muscle tissue these bubbles cause paralysis and block veins and arteries like blood clots. Unconsciousness and death may result.

Coming up from the depths slowly and by degrees prevents this attack. This slows down the rate at which nitrogen leaves the tissues until the body is prepared to receive the normal atmospheric pressure.

HOW DOES BAKING POWDER MAKE THE DOUGH RISE?

All baking powders contain baking soda (sodium bicarbonate), starch, and a third substance, either cream of tartar or phosphate. When the baking powder is mixed into the flour, and a little water is added to make the dough, the phosphate or cream of tartar in the baking powder reacts with the baking soda to set free bubbles of carbon dioxide gas. When covered and set aside for an hour or two, the dough rises to almost twice its size, due to the released gas. It is then set in the oven to bake.

CAN SCIENCE DUPLICATE A HUMAN CHEMICAL?

In 1828 a German chemist, Dr. Friedrich Wohler, synthetically produced urea in the laboratory. This is a chemical found in the urine, blood and lymph of mammals, including man. It is now manufactured on a large scale from ammonia and carbon dioxide and is important in the manufacture of plastics.

Recently two scientists put together the hormone known as ACTH. It is normally made by the pituitary gland in the brain and stimulates the activity of the adrenal glands and several others. This breakthrough is important in future medical treatment and research, because the way is now open to reproduce other natural proteins in medical therapy and research.

At present, extracts of ACTH from pig and sheep pituitary glands are sometimes used in treatment of severe rheumatic, allergic and certain other disorders. The American Chemical Society said, "Now molecules can be made in a space of a few weeks or months that previously might have taken a portion of a lifetime."

Wouldn't it be wonderful if some day, man could duplicate blood plasma?

WHAT PURPOSE DO TEARS SERVE?

A famous criminal lawyer used to create a flow of tears to make the jury cry. Seldom did they find his defendant guilty. You might justifiably say that the purpose of his tears was quite unusual. Now let us examine the real function of tears.

There are two tear glands in each eye, located under the eyelid over the eye. They secrete salty tears which come out through several small ducts in the underside of the lid. These tears wash the eyeball and carry away foreign particles, then movc toward the ducts that lead into the nose. In this process the eyeball is also lubricated to prevent infection, because a dry cornea could lead to blindness. With each eyeblink, a little fluid is sucked from the gland. But under strong emotional feeling, such as in grief, anger or hearty laughter, the suction is so strong that the gland muscles tighten up and squeeze out more tears than the ducts can carry. It is then that the overflow comes.

Tears contain not only salt and substances that fight off bacteria, but also proteins that make the eyes immune to infection. Thus you could say that the lawyer cried to make his eyes immune to infection, and his clients immune to punishment.

WHICH PLANTS RESTORE THE EARTH'S MINERALS?

It's always the same old story. Easy to borrow and hard to return. How well the farmer knows this. Each year he must try to return to the soil the minerals his crop took out. Unless he puts them back his crop will grow smaller and smaller. The one plant that is most ideal for restoring nitrogen to the earth is alfalfa. Not only is it hardy and able to survive dry climate and long periods of drought, but its roots can penetrate 17 feet into the ground to reach moist earth and water. With farmers it is popular because it is grown as food for animals. Its most important quality is its ability to take nitrogen gas from the air and return it to the earth in the form of solid nitrates. That is why it is one of the chief plants used by farmers when they alternate their crops.

Two plants of the pea family, peanuts and clover, do the same thing. Again, one must wonder at nature's beautiful mystery, to be able to change a gas into a most essential solid in our chain of life.

HOW LONG CAN BACTERIA LIVE?

Bacteria that lived 600 million years ago were extracted from salt layers in 1962. A more recent discovery of live bacteria was made in northwest England. They came from mud at the bottom of a lake, and by radio-carbon dating were found to be 1,500 years old. They were taken in a core one meter long at a depth of 60 meters.

WHAT ARE SOME EXAMPLES OF PROTECTIVE COLORATION?

Many animals are colored in ways to conceal them from their enemies. The polar bear is hard to see in the snow, the copperhead snake looks like the dead leaves it hides in, and the chameleons, flounders, octopus, squid and cuttlefish change their colors at will for protection and advantage in attack. The praying mantis looks like a twig, butterflies resemble leaves and flowers, and many birds are colored like their surroundings.

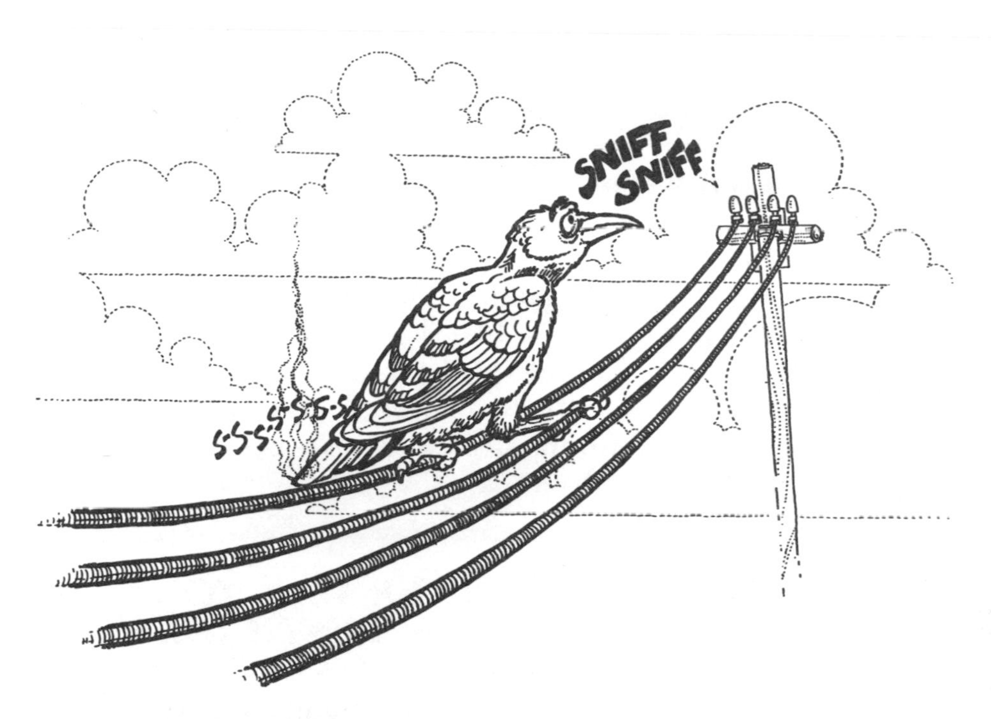

WHY AREN'T BIRDS ELECTROCUTED ON ELECTRIC WIRES?

For a bird to get electrocuted, its body must be grounded; that is, there must be a completed circuit in order for the current to go through its body. If the bird could stand with one leg on the wire and one on the ground, the circuit would be completed. In all cases where a person has been electrocuted, part of the body touched the wire and another part touched an uninsulated object, such as the ground or something touching the ground.

WHERE DOES THE RAINBOW GET ITS SEVEN BASIC COLORS?

When white light like the rays of the sun passes through a prism (a triangular piece of glass) or, as in the case of the rainbow, through little drops of rain, the white rays of light separate and break into the seven colors of the spectrum. They are: violet, indigo, blue, green, yellow, orange, and red. This is called dispersion. Each color travels at a different rate of speed through the raindrops and is therefore bent or dispersed at different angles. Therefore you can see the seven colors and also a blend of most of them. Now you know that the beautiful band of colors called a rainbow usually follows a rainfall and is really the dispersion of sunlight into spectrum colors as they pass through the raindrops.

IS THERE A BETTER WAY TO FIGHT BUGS THAN PESTICIDES?

Evidently there is a better way. Some scientists are presently working on a method to inhibit a mosquito's ability to carry malaria by means of an enzyme or protein. In certain southern states, traps were baited with chemicals to draw sex-crazed boll weevils by the thousands. The bait was the chemical which weevils produce to attract mates. Elsewhere, a gypsy moth sex attractant was sprayed into the air that drove the moths into such wild confusion that their sex drive was completely disabled.

Prof. James Babler, a chemist at Loyola University, says, "My job is to find the right chemical to drive an insect buggy."

The process is long and tedious. Collecting and squashing hundreds of thousands of insects is only part of it. Separating the thousands of different chemicals that make up the bug is the other part of the job. The tedious process takes years because only minute amounts of chemicals can be obtained. They are presently on the verge of perfecting the right chemical for cockroaches, but it has taken years. The researchers' job is to find a sex lure for mosquitos to replace the repellents and poison sprays.

WHAT DO WE MEAN BY REGENERATION?

Regeneration in plants and animals means the power to replace a lost or damaged part. We all know that many plants, trees and shrubs, when cut low to the ground, will send up new shoots. But how about animals?

The lower forms of animals such as sponges, coelenterates and simple worms, possess a great power of regeneration. In fact, a starfish may lose one of its arms and regenerate it. It has been known to survive being cut in half and return as two starfish. An earthworm, when cut in half, will grow a new tail on the head end. A flatworm will grow a new head. Many insects, before full growth, can regenerate a lost leg. A fiddler crab, in an effort to escape an enemy, will actually throw off his large claw, and the remaining small claw will grow into a large claw. A new small claw grows in the place of the departed large claw. A lizard called a "glass

snake" escapes its enemies by breaking off the end of its tail; it later grows a new one.

Now how about us, the higher mammals? Unfortunately, we are only capable of regenerating hair, nails, skin and liver tissue.

It's interesting to speculate what tricks nature could play on us if we had the power to regenerate a lost organ such as a leg or a nose. What if it grew back twice as long as the original?

WHO ARE THE ARMY ANTS?

These fearless ants live in Africa and South Central America. Although they are blind, they fear nothing, and live entirely on other insects and larger animals.

Without a permanent home or nest, they spend their whole lives roaming in large bands through jungles and villages in search of prey. On these hunting expeditions they carry their young and their queen ant with them.

In attack they are so swift and fierce that every living creature that knows about them fears them. They will eat every living creature, large or small, that they encounter and catch. In a village they will devour all the poultry and pets they can find. When they leave a house, there is not a living thing in it. Rats, snakes, roaches, and all other pests have either run away or fallen victim to their attack.

It's a wonder that some military force of the world hasn't put to use these army ants as an advance unit in a military attack. Could be they're afraid of an army ant rebellion.

HOW DOES THE SHAPE OF THE WINGS HELP A PLANE?

In order for a plane to fly, it must create two forces called thrust and lift. Thrust, the forward motion, is supplied by the propeller or the hot escaping gases from the rear of a jet engine. To create an upward lift that will overcome the downward force of gravity, the plane must be moving speedily with wings that have their top surface curved and their bottom surface flat. This is based on the principle that air will pass more swiftly over a curved surface than a flat one. In this way, two different air pressures are created, with the greater pressure under the wing. The difference is what gives the plane the upward lift and helps support the plane in the air.

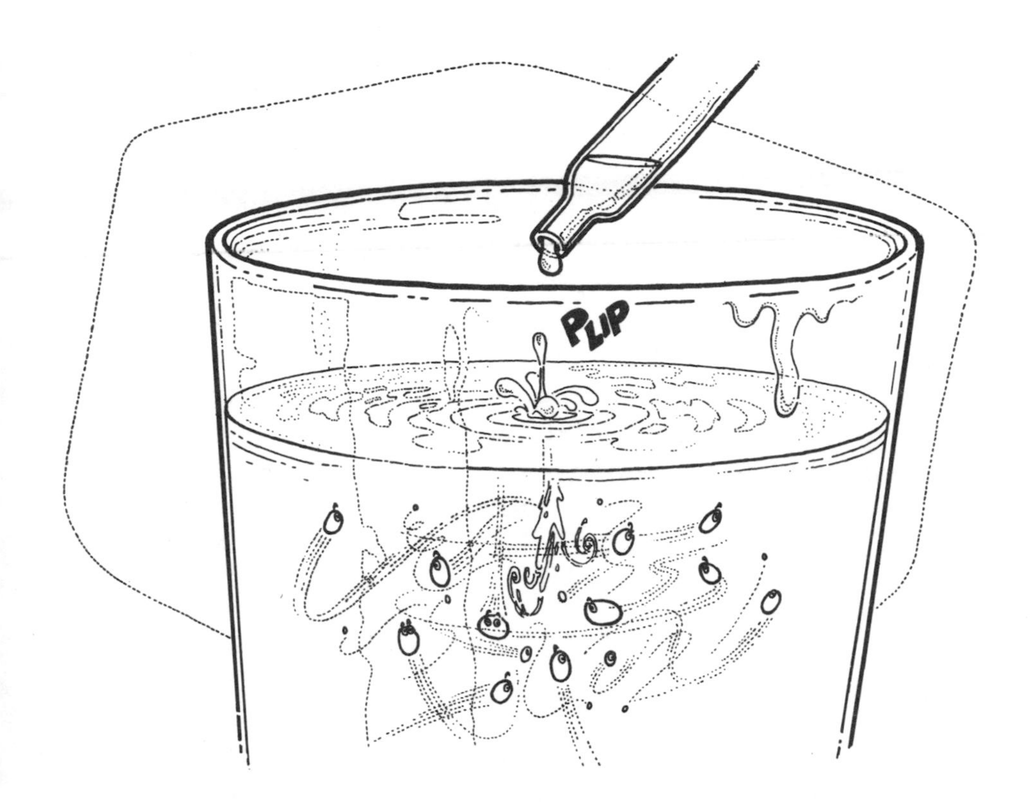

CAN YOU PROVE MOLECULES ARE ALWAYS IN MOTION?

Yes, in a very simple experiment. Gently place a drop of blue ink on the surface of a glass of water. Use an eyedropper to make sure it is done gently. In a few moments, before your very eyes, the single drop of ink will become diffused, and will disappear into the water, turning the entire glass of water a dilute blue. The ink and the water are both composed of millions of ultramicroscopic molecules. Since the molecules in both substances are in constant motion, they continue to bump each other till the inkdrop commingles and disappears among the water molecules, right down to the very bottom of the glass.

DO UNDERWATER CREATURES COMMUNICATE?

Yes. With the help of a team of sea scientists, two killer whales in captivity talked to each other from one city to another over long distance telephone lines using hydrophones at each end as the receiver-transmitter.

Although we do not understand the sounds they give out, it is certain that undersea mammals such as the porpoises understand each other. Extensive recordings have been made of many whale sounds, and those of the drumfish, groupers, sea lions, seals, puffers, crabs, snapping shrimp and lobsters. All of these experiments lead us to believe that they must have a means of communication with one another, and these different sounds are the messages they convey.

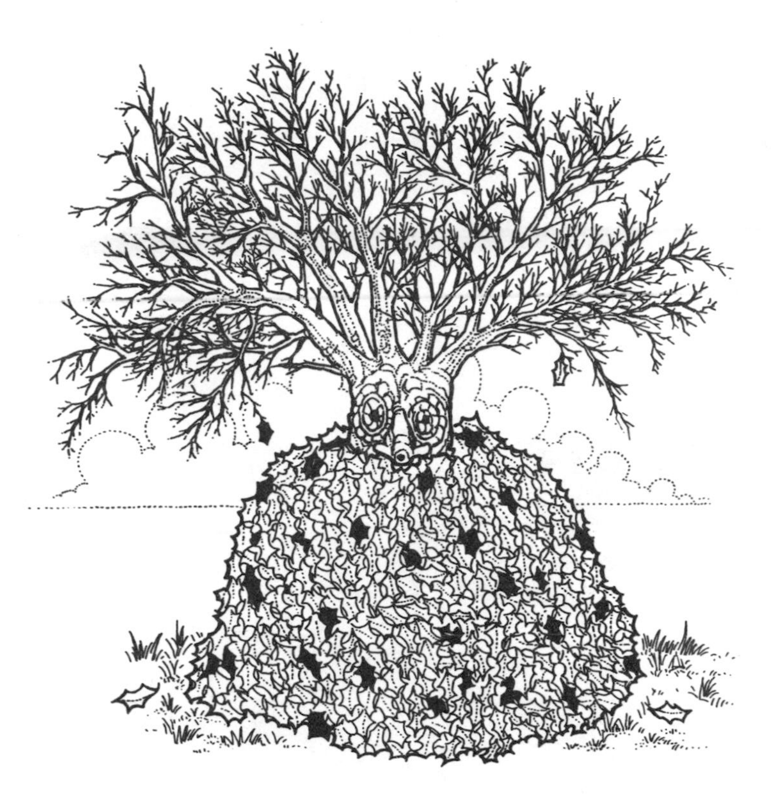

WHY DO MOST TREES SHED THEIR LEAVES IN THE FALL?

If your way of life depended upon the supply of water that you needed to run your home, and if every now and then your water supply were shut off, it wouldn't be long before you would move to another place where the supply was plentiful. But a tree cannot get up and move away when the water supply gets low in the wintertime. So nature has found a way for the tree to survive through the winter. In the fall, most trees shed their leaves in order to survive the winter when water gets scarce. Less rain and more freezing temperatures bring this about. In the fall, the color variations in the leaves are due to the variation of chlorophyll in them. It takes sunshine to make the chlorophyll that gives the leaves their rich green color. In the fall, with less sunshine, the leaves change from a rich green to the various red and golden shades they acquire.

WHAT HAPPENS WHEN YOU MAKE HARD WATER SOFT?

The "hard" carbonates in the water are due to the presence of dissolved mineral materials, such as limestone, at the source. To "soften" the water means to eliminate the dissolved limestone. So a chemical is put into the water which combines with the limestone and invisibly takes it out of the water.

WHAT IS HIBERNATION AND HOW DO THE ANIMALS SURVIVE?

The word hibernate comes from the Latin, meaning winter sleep. Animals grow fatter before hibernating time and also slower in movement as the cold begins to set in. Some animals, like the chipmunk and gopher, store food which they eat during breaks in their sleep, and then return to sleep again. Animals that don't hide food for future use do not starve during hibernation, because they use up the fat stored in their bodies.

Some animals lose about a third of their weight during this period of sleep. Their breathing is very slow, their heartbeat scarcely discernible, and their body temperature about as low as their surroundings. An increase in temperature and pangs of hunger will awaken them.

The period of sleep varies with the animal and the circumstances, but the champion of all winter sleepers is the common dormouse, who hibernates for five or six months from October to April.

Most all burrowing animals hibernate. Among hibernators you will find the bear, woodchuck, frog, snail, snake, skunk, squirrel and bat. A few butterflies, bees and flies also hibernate.

WHY DO SOME PEOPLE SNORE?

Did you ever blow air across a blade of grass and get a vibrating sound? Well, from the palate in your mouth right above the base of the tongue there is a soft mass of tissue called the uvula, which hangs down and acts like that blade of grass. The American Medical Association estimated that men and women share the record equally among the 25 million Americans who snore. These include children who have enlarged adenoids or tonsils, and persons with a blocked nose due to a bent nasal bone. Here, minor surgery could help the snorer. Why do some snore softly while others are unbearably loud? Well, the number of vibrations depends on the size, thickness, and flexibility of the tissues and the force of the flow of air.

WHY DOES SAND COME IN DIFFERENT COLORS?

Sand is chiefly composed of quartz, feldspar and bits of mica. The erosion of exposed granite rock in mountains gradually broke the granite into small sand particles that were washed down by rain. Other beach sands called basalt rock were produced by volcanic ash, originally black lava, as in Hawaii, and bits of broken shells and coral material, as in Florida and Bermuda. So you see, white sand came from coral rock, pink sand from masses of granulated pink seashells, black sand from volcanic ash, and multicolored sand from a mixture of ground multicolored shells.

Sand has many commercial uses, such as the making of mortar and concrete and the manufacture of glass.

WHY CAN'T ASHES BURN?

Substances which are already oxidized (combined with oxygen) can no longer take in oxygen or be oxidized again. While a material is burning, it is being changed into something else. That something is ashes. This new substance, the leftovers of the fire called ashes, no longer has anything in it to unite with oxygen. Therefore it cannot burn again.

WHY DOES PAINT PEEL?

A painted surface unites with the oxygen in the air in a process called oxidation. It is a process of fading, aging, and turning into another substance like iron into rust. While this process goes on, the outside layer of the paint becomes very dry and hard, and while doing so, begins to shrink. Thus, while shrinking, it pulls itself away from the wall. Wallpaper does the same because the paste dries up, shrinks, and pulls away from the wall.

ARE SPONGES PLANTS OR ANIMALS?

At one time sponges were thought to be plants because they did not move around as most animals do, and because most sponges are branched and look like plants. But they are a group of sea animals that make up the next to the lowest main division or phylum. Their colors vary from black to blue, green, red, purple and yellow. Their shapes vary depending upon their surroundings, and they live in abundance from the shallow waters to the deep seas.

Starting out as a single cell, the sponge divides and multiplies itself until it becomes a little larva covered with hair. Then it swims to sea, settles to the bottom, attaches itself to a rock by means of its base or stem, and begins to develop into an adult sponge. They can range in size from a quarter inch to five or six feet in height.

Remains of sponges have been found in the oldest rocks that contain fossils, proving their ancestry existed from earliest times. Commercial sponges grow only in warm waters and come mainly from the eastern part of the Mediterranean Sea.

WHEN HEATING A ROD, WHY DOES THE OTHER END GET HOT?

Molecules of matter are continuously in motion. Heating one end of a rod drives the molecules into greater motion, and the friction of the molecules bouncing around in greater motion finally heats the other end. The process of heat passing through a solid is called conduction.

WHAT CAUSES BALDNESS?

Not so long ago one type of temporary baldness was caused by the male parent. On the first day of vacation, when school was closed for the summer, Father took the boy to the barber shop and said, "Give him a haircut to last all summer."

Other causes of baldness are infection, disease, poor gland function, metal poisoning, and dandruff which kills hair roots. But if the head of hair is strong and healthy, it can support a weight of 2,000 pounds. Hair will live two to four years and is replaced when it falls out, even after the prime of life.

Another cause of baldness is the inheritance factor, which affects males more than females. While the women of a bald family are somewhat liable to baldness, they are more likely to retain their hair than are men. But they pass on the inherited trait towards baldness to their sons. This trait could even determine the pattern of baldness.

WHY DOES WATER BOIL AT LOWER TEMPERATURES ON A MOUNTAIN?

At sea level, where air pressure is 14.7 pounds per square inch, water will boil at 100 degrees Celsius. But since air pressure decreases the higher we rise in the atmosphere, and since mountain pressure is lower than sea level pressure, it doesn't take as many degrees of temperature to make the water boil. So if you want a quick cup of hot tea, just climb a mountain and make it.

HOW MANY TYPES OF ENERGY ARE THERE?

Here are some examples.

1. *Kinetic energy.* Wind or falling water from a dam.
2. *Mechanical energy.* A wheel turned by falling water.
3. *Heat energy.* Results from anything burning.
4. *Chemical energy.* Two substances changing into a new substance.
5. *Electrical energy.* Obtained from an electric current.
6. *Solar energy.* Heat from the sun.
7. *Light energy.* Light from the sun or other source.
8. *Potential energy.* Energy locked up in coal, wood, oil, gas and food.

9. *Muscular energy.* Energy from muscular effort.

10. *Steam energy.* Energy derived from steam pressure.

11. *Atomic or nuclear energy.* From the splitting atom.

12. *Gravitational energy.* From anything that falls.

WHERE DO SEASHELLS GET THEIR PRETTY COLORS?

About nineteen different minerals have been extracted from our waters, and each one, alone or in combination, can produce a different color in the substance subjected to its chemical influence. An example is the orange-brown of iron oxides found abundantly in beautifully colored seashells. There is a hereditary mechanism for acquiring pigments from the environment of the ocean, and using them in the creature's body to produce color. This hereditary mechanism is the same in like species, which is why fishes and shellfish of the same species are colored alike.

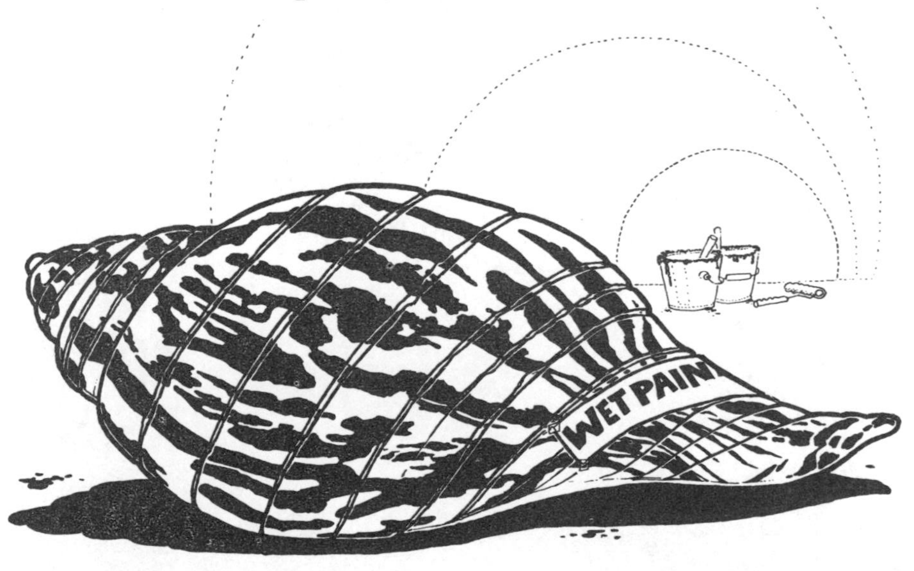

WHAT'S THE DIFFERENCE BETWEEN A SOLAR AND A LUNAR ECLIPSE?

The earth revolves around the sun, and the moon revolves around the earth, each in its separate orbit. At one time or another, when the orbits of the moon and the earth are in the same line, either the moon or the sun gets blocked out. In a solar eclipse the moon is between the earth and the sun. In a lunar eclipse the earth is between the moon and the sun. In a solar eclipse the moon blocks out the sun, while in a lunar eclipse the earth blocks out the moon.

WHAT WAS MANKIND'S GREATEST DEVELOPMENT?

The answer may be controversial, but most scientists will agree it was the development of languages. Without the ability to communicate, and record for future generations the progress of the times, no invention would be possible.

HOW DOES A RATTLESNAKE SENSE ITS PREY?

There are about thirty kinds of these very poisonous snakes in the United States. Since they have neither an inner nor an outer ear, they have no sense of hearing whatsoever. But nature, to compensate, has endowed them with another device to sense an approaching enemy or prey. On their forked tongues are Hudsonian cells which are highly sensitive to heat. By constantly sticking out its tongue the snake can detect the proximity of an approaching body. This device, coupled with the ability to follow a moving object with its eyes, seems to make up amply for any hearing deficiency.

WHY IS OUR BLOOD SALTY?

There really is no scientific answer to this question. It is only a theory and comes from Darwin's theory of evolution, which says that we were animals of the sea millions of years ago. Our blood became salty from the sea and ocean form of life. Millions of years after our ancestors left the seas and came out to live on land, their blood still carried the salt content, and continued to pass it along. What is most amazing and interesting is the fact that there is almost the same proportion of minerals in the seas as there is in our body.

HOW MANY BLOOD COLORS ARE THERE?

There are three blood colors. Mammals have red blood, lobsters have blue blood, and insects have yellow blood.

HOW DOES THE X-RAY MACHINE WORK?

X-ray tubes, which are really modified electric bulbs with a different internal structure, contain a vacuum which is about one billionth of an atmosphere.

Inside the tube at the threaded end is a coil of tungsten wire which is heated by a current going through it. The heated tungsten coil is the cathode or the negative terminal of a high-voltage circuit. When the current is turned on, the negative charges, called electrons, shoot out of the tungsten coil with tremendous force against the oblique face of the anticathode or anode which faces it a small distance away. This oblique end of the anode in the tube is also made of tungsten, and the striking of the electrons with tremendous speed against the oblique face of the anode causes it to radiate X-rays.

The X-rays pass through the thing or person to be examined but are blocked by denser materials. Since the less dense materials allow the rays to go through, a light and dark picture of inner structures can be obtained.

HOW CAN YOU DETECT A POISONOUS MUSHROOM?

The answer is that you cannot detect a poisonous mushroom until you eat it. But what fool wants to make the test? So the best advice is never to pick and eat anything that remotely resembles a toadstool or a mushroom. Of course, there are experts who can tell the difference. You must therefore rely upon the market you deal with, and trust that they make their purchases from people who raise and cultivate good edible mushrooms.

IS "LEAD PENCIL" A MISNOMER?

When you speak of a lead pencil you are talking about pieces of lead used for pencils by the ancient Egyptians and Romans. They never dreamed that graphite could make a better pencil because of its softness and because it can write darker. Because it is a mineral of natural carbon, it can even be made artificially. It's been thousands of years since lead was used, so we really should call it a graphite pencil.

To make a modern pencil, ground graphite is mixed with clay and shaped into long thin rods. More clay gives a hard pencil and more graphite a soft one. The soft rods, which come out of a pressure machine, are then cut, dried and baked in an oven to be made ready for their wooden cases. Then the rods are set into the grooves of half a block of wood containing about six grooves made to fit the rods. The grooved block is coated with glue before the rods are set in and the other half of the block is placed on the first half and put under pressure until the glue dries. When the blocks are dry, they are cut into six pencils each.

HOW DO BEES BUILD A HONEYCOMB?

When beehives get overcrowded, a large number of bees that may run into the thousands leaves the hive. This great mass, crowding around a queen bee in the center, soon becomes a cloud ten to twenty feet in diameter. It is now what you call a swarm of bees. They will finally settle in a natural hollow in wood or rock and start to build a new home.

Soon some fancy acrobatics take place. A group of bees begins clinging to the ceiling of their new hive, and the other bees start to hang down from them to form long chains. For a day and a night they remain suspended in this position, during which time some internal wax-making chemistry takes place. It isn't long before the wax begins to appear on the wax plates located on their abdomens. While all this is going on, other bees are feeding them honey. As soon as the wax appears, worker bees scrape it off the plates and chew it well before they apply it to the comb being shaped in the hive.

The combs are built from the top downward in an exact pattern. Both the honeycombs and the brood combs for the eggs are flat. They consist of two layers of six-sided cells placed end to end, with a thin sheet of wax between them. The cell walls are thin and light, but they can hold many times their own weight in honey.

It has been roughly estimated that the bees in a hive travel about fifty thousand miles, or about twice the distance around the earth, just to produce a pound of honey.

WHAT VEGETABLES WERE UNKNOWN TO OUR FOREFATHERS?

Artichoke. The globe artichoke is a native of the Mediterranean region and is now grown in California. The Jerusalem artichoke is cultivated in France.

Soybean. This was first grown in the Orient and brought to the Western world during World War Two.

Swiss Chard is a variety of the beet plant and was grown as far back as 350 B.C. It was brought from Switzerland to the United States in 1806 and much of it is now grown in Massachusetts.

Chinese Cabbage, related to mustard and cabbage, is one of the oldest food crops of China and is now grown in the United States.

Cowpea or Black-eyed Pea grows wild in Asia and is now grown in the United States chiefly for forage.

Broccoli, which belongs to the cabbage family, first came from southern Europe and is now grown along the seacoasts and in the Great Lakes region.

French Endive is a leafy vegetable related to chicory and grows wild in the East Indies, from whence it came to other parts of the world.

Kohlrabi belongs to the cabbage family and tastes like rutabaga. It was raised in Europe before Christian times and is now grown mostly in home gardens.

Zucchini or Italian Squash came from Italy and is now a very popular vegetable.

WHAT WOULD HAPPEN IF ALL THE ICE IN THE WORLD MELTED?

This event is extremely remote, but if it did happen, it would be measured in thousands of years. It is a highly speculative situation, and it is just as pleasant to think that the increased weight of the water would cause the ocean floors to sink and the land masses to rise as it is to think just the opposite.

But if the ice should suddenly melt, sea level all over the world could rise 500 or 600 feet. There is also the possibility that the ice age is not yet over, and that the ice caps may again increase in size. If another glacial advance like the last one should occur, important areas of the world would be covered, and forced migrations would be widespread. But remember again that the ocean floor could sink, and new land masses arise.

WHY IS THE PITUITARY GLAND SO IMPORTANT?

Picture, if you please, all the endocrine glands in an argument over which one had the most important function in the human body. The thyroid has just finished his self-appraisal with a self-serving speech, and the pituitary arises to speak. This is what he says:

"See here, you fellow workers, I'm the most important gland around here. I may only be the size of an acorn, but without the many hormones made by my anterior and posterior lobes, I doubt whether most of you could function properly. I hang down from the bottom of the brain just above the nasal passage, and I know what's going on all the time. If I were to get sick and take a couple of days off, or if by accident I should stop functioning altogether, most of you would be in a lot of trouble.

"I help regulate the normal function of the thyroid, which is important to the proper growth and development of the skeletal system.

"I control the normal output of the sex glands, and have a controlling effect on the pancreas, which as you know is concerned with the body's use of sugar.

"I control the adrenal glands, which regulate the salt and water levels and also secrete stimulants.

"I control the contraction of the walls of the uterus, I raise blood pressure, I stimulate intestinal peristalsis, and I promote the retention of water in the body by the kidney tubules.

"There are a lot more things that I believe I do, but I cannot tell you about them until the scientists find out what they are.

"Now do you see how important I am?"

WHY WAS CAPE HATTERAS THE GRAVEYARD OF SHIPS?

Cape Hatteras is at the tip of a long chain of sand bars and low islands that lie off the coast of North Carolina. Here the Gulf Stream flows about twenty miles east of the cape. In the past, when southbound vessels were driven too near the coast they were wrecked on the cape. Because of the many ships and lives that were lost here, it became known throughout the world as the graveyard of ships.

Its reputation comes from a combination of factors which make it a dangerous area. It gets sudden storms of great intensity without warning, it has constantly shifting sand bars hard to detect, and at times has very strong currents of great velocity, or no current at all, or a current that reverses its direction.

Today, with a lighthouse built at Cape Hatteras at the request and with the influence of Alexander Hamilton, and with modern electronic aids to navigation, Cape Hatteras has lost much of its fear and dread for mariners.

WHAT CAUSES NIGHTMARES?

Everyone will give you a different reason. Parents may have nightmares because a child is using drugs, or a son can't resist pilfering from his boss, or because daughters are going out with far-out boyfriends.

Now, generally, you get the idea. Nightmares are

related to a worrisome or pressing problem. Of course these dreams filled with terror could be brought on by other things, such as horror movies, terror TV films, drugs, even those prescribed, or the occurrence of a frightening experience not long before bedtime. With children this is usually the case. They awaken screaming because of over-excitement or the watching of a dread-filled movie just before bedtime.

Other causes of nightmares could be indigestion, poor blood circulation or restricted breathing. But all in all, don't take nightmares too seriously, as they are only unpleasant dreams and nothing more.

WHO ARE THE MIDGET MONSTERS OF THE SEA?

The first of these is the deep sea angler fish, which gets its prey by attracting it with a light on its head.

The *Linophryne arborifer* has a light on its head and a luminous beard, both of which attract its prey.

The *Photocorynus siniceps,* though only 2½ inches in size, is a fierce deep-sea prowler. The male is only ⅖ of an inch long.

The dragon fish is another one of the ferocious undersea monsters.

Although their growth is stunted because of the scarcity of food in the dark cold world of the oceans a mile below, they could be compared in ferocity to the tiny unbeatable little shrew on land. The water in which they fight for existence reaches a pressure of 2,000 pounds per square inch, and any one of these monsters would quickly burst if brought to the surface.

HOW DOES A BEE MAKE HONEY?

The tongue of the worker bee is a long slender underlip rolled into a tube. This tube the bee inserts into the liquid called nectar, located at the base of the flower petals. It then sips all it can until its special stomach, called a honey sac, is full. This special stomach is really an enlargement of the esophagus. Here is where the nectar goes through its first series of changes which turn it into honey.

When the bee returns to its hive, it places the nectar in storage, where the final change into honey takes place.

When a bee leaves a hive in search of food, it seems it is able to retain a knowledge of the direction by means of polarized light from the sky which strikes its eyes.

At the beehive the bee is able to communicate to the other bees, in a fantastic way, not only that it has found a source of nectar, but the direction and approximate distance from the hive to the flowers. With the true spirit of sharing in the family of bees, a bee need never worry that he will go without food.

HOW LONG CAN VEGETABLES BE STORED?

Different fruits and vegetables have varying storage periods before they start to ripen and deteriorate. Because the extension of the storage period can be a boon to mankind, scientists are ever striving to discover ways of extending them.

Bountiful seasons could take care of seasons of drought, and fruits and vegetables being always at hand and plentiful could help supply a needy nation. The potato is a good example of such progress.

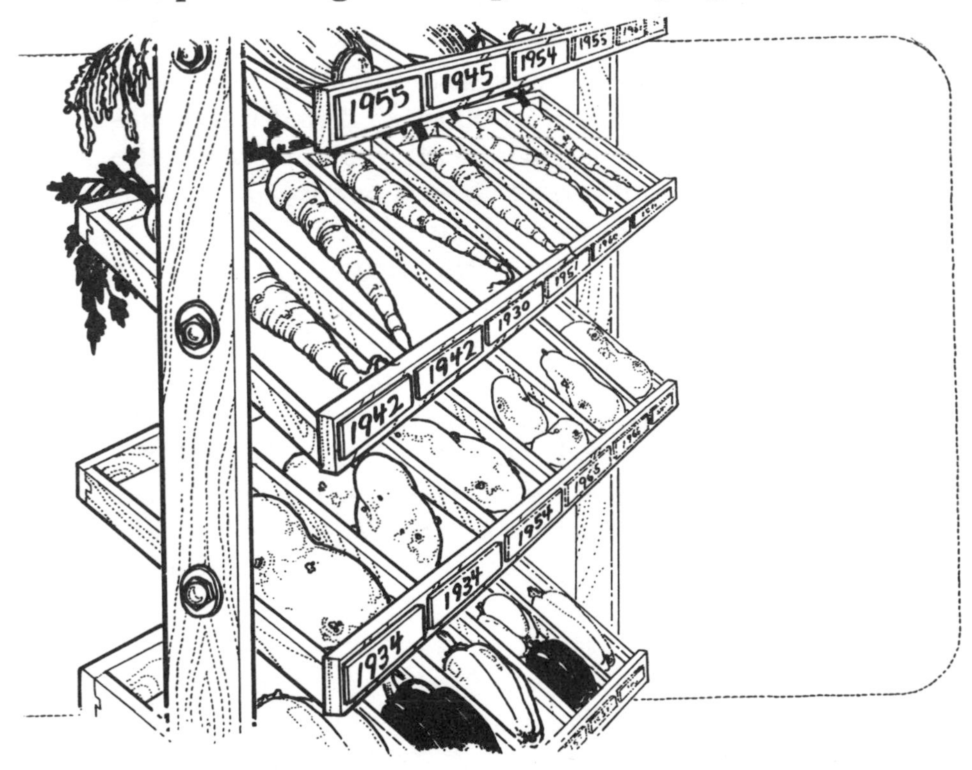

The potato tuber is really a modified stem and the eyes are the buds. A freshly dug potato at one time had a storage life of only a few months before the eyes began to sprout. Now, by the application of a synthetic auxin spray, the storage life has been increased to as much as three years.

With the ability and capacity to store the potato, we need never fear a shortage.

IS THE STORY OF ATLANTIS A MYTH?

A legend is a legend is a legend. But since over 5,000 learned people have belabored the subject of the lost city of Atlantis, there may be a lot more to the story than just myth. If it is still hard to believe that such a city existed, the same doubts existed for hundreds and hundreds of years about the cities of Troy, Pompeii and Herculaneum, until they were discovered and dug up. Who knows? Someday someone will make a discovery of the lost city of Atlantis, and prove Plato's account of the city was true.

This city of supposedly great civilization was so far ahead of its time that by 10,000 B.C. it had built temples, ships and canals. It had almost conquered the small world about it except Greece, when suddenly it was engulfed by the sea and disappeared. This overnight disappearance of Atlantis without a trace gave rise to all the later historical speculations about it.

Plato thought Atlantis was west of the Pillars of Hercules, but since he wasn't too specific about where the Pillars were located, some assume that he meant the Straits of Gibraltar and others the Strait of Messina between Italy and Sicily. But wherever it is, some day a deep-sea diver who specializes in the recovery of lost cities may come up from his dive and say "Eureka, I found Atlantis."

WHY DOES A KITE NEED A TAIL?

The tail of a kite is much like the stabilizer of a plane. It acts as a stabilizer to keep the kite from rolling and spinning to the ground.

WHY DO SOME PEOPLE TAN WHILE OTHERS BURN?

Sunburn is an inflammation of the skin that comes from overexposure to the ultraviolet rays of the sun. There is a dark pigment in our skin called melanin which is made by pigment cells called melanocytes. When the skin is exposed to the ultraviolet rays, this pigment absorbs them. The cells produce the pigment melanin at rates which speed up on exposure to sunlight. When they speed up the manufacture of melanin, the skin becomes darker. This gives one a tan. But if the sun lovers remain in the sun too long, that is, to the point where the cells no longer can produce the pigment, then the excess ultraviolet sun rays cause certain unknown chemicals in the skin and blood vessels to swell the vessels and make the skin red and sore. This is called a burn and can sometimes be so severe as to require hospital attention. The rate of pigment production varies in people, and that's why some can take a little, and some can take a lot of sun.

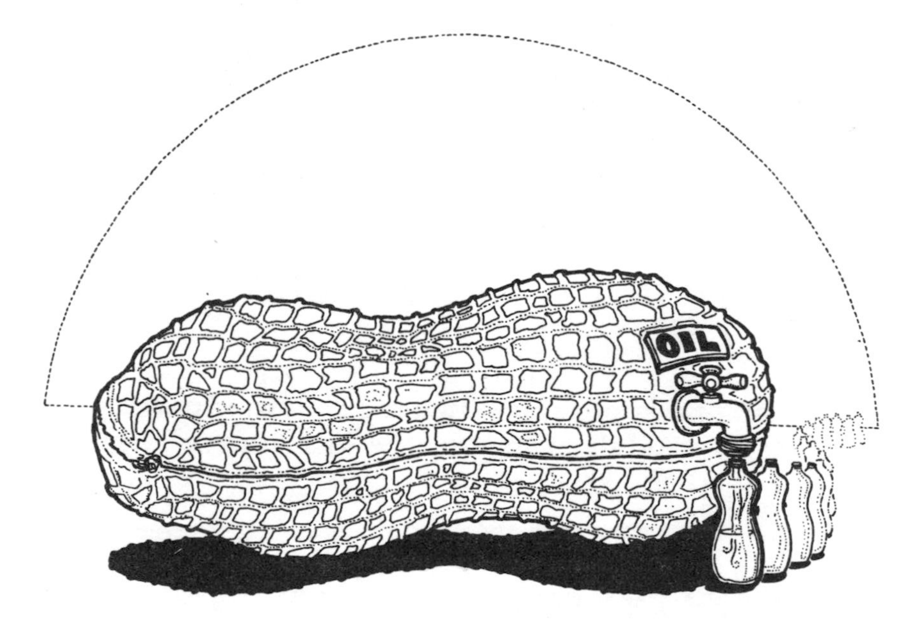

WHERE DOES A PEANUT GET ITS OIL?

By the process known as photosynthesis, a plant uses the dissolved minerals from the ground. The dissolved minerals, by the process of osmosis, pass into the internal structure of the plant by going into the root hairs, through the roots, and into the plant proper. There the plant factory first makes sugar, which in time turns into proteins, fats, and oils. Each plant, according to the genetic structure of its seeds, manufactures the particular substance of its species. Thus, the peanut plant ends up with oil.

This oil is used to make margarine, soap, glycerin, medicines, massage oil, salad oil, and packing oil, and is used in the kitchens of homes and restaurants for deep-fat cooking.

Exactly how the change takes place from the raw dissolved minerals, absorbed from the earth, into the finished peanut oil in the plant, still preoccupies the studies of the chemist. Gradually it is hoped they will solve this mystery.

HOW DO HONEY ANTS GET THEIR NAME?

In the southwestern part of the United States, there live certain ants whose main food is honey. This they get from the galls or abnormal growths of oak trees. Some of the worker ants turn themselves into storage tanks by receiving the honey from the other ants and storing it in their stomachs.

Soon their abdomens grow so big and round that they look like small marbles. They become so fat that they can hardly move. So they spend the rest of their lives hanging by their feet from the ceilings of special rooms in the nest. When their hungry fellow workers ask for food, the storage ants give them a little of their honey.

In a large nest there may be as many as 300 ants filled with honey. In Mexico, it is said that the country people sometimes open the nests and eat the honey-filled ants.

HOW MUCH HELP CAN A LEVER GIVE?

A lever is a bar that rests on a fixed part called a fulcrum, and is used to lift a great weight at one end with small effort at the other end.

If the lever is 10 feet long and a 1,000-pound safe rests at one end, one foot from the fulcrum, a man could raise that safe with a force of 100 pounds. If the lever were only 5 feet long, he would have to use a force of 200 pounds. But if the lever were 20 feet long, he could raise the safe with a force of only 50 pounds. If the lever were 100 feet long, he could raise that safe with only 10 pounds of effort.

Without going into the mathematics of levers, you can see how correct Archimedes was when he said, "Give me a place to stand, a lever long enough, and I will be able to move the world." And this was in 240 B.C.

WHAT IS MOLTING AND WHO DOES IT?

If you could look into a mirror, see your wrinkles and say, "I think I'll shed my skin today," wouldn't that be something? All those useless oils, youth restorers, wrinkle fighters and fake skin tighteners would go out of business at once. And for no cost at all, you would have a complete new, youthful-looking skin. But unfortunately, only a few species in the animal kingdom are capable of skin-shedding or molting. The lemming and ptarmigan, unlike us, save manicure expenses by shedding their old claws for new ones. Snakes and lizards have long been known to shed their skins. Spiders, shellfish, lobsters and crabs also drop their coverings to replace them with new ones.

Every bird sheds its feathers at least once each year.

Some do it three times a year, but the feathers do not come out all at the same time. They fall out evenly on each side, beginning with the neck down, and are immediately replaced in correspondingly regular order with new feathers. Once a feather is full grown, it begins to die. Blood vessels and nerves are not involved, and it takes from four to six weeks for a complete molting to take place.

WHY DO CATS LIKE CATNIP SO MUCH?

Catnip or catmint is a plant of the mint family that grows about two or three feet tall and has little clusters of white and pale-blue flowers. This North American and European roadside weed for some reason has a stimulating effect on cats like a mild narcotic. Cats love to roll and play in the leaves and occasionally eat them. Perhaps as a tribute to the cat, the plant got its name.

Today, lovers of kittens and cats can buy the dried leaves and stems crushed and packed into small cloth bags for their pets.

WHAT MAKES A JUMPING BEAN JUMP?

While we call them Mexican jumping beans, in Mexico they are called "los brincadores," the leapers. About thirty million of these beans find their way each year into the United States, Japan and Europe, to be sold as items of unique interest. In reality, it is a seed from a shrub of the spurge family, inside of which resides a larval insect. Some time before, a moth laid an egg in the flower of the shrub. Now, in trying to escape from its prison, the larva grasps the inside wall of the pod with its legs and recoils its body, thus giving to the bean its jumping motion.

CAN CHICKENS BE INDUCED TO LAY MORE EGGS?

They certainly can, but it has to be by trickery. How would you like to have your boss turn the clock back on you while you were out on your coffee break? How would you like to put in extra time each day by such deception? Well, if you found out about it, you'd be justified in suing for the extra back pay due you. But to whom can the poor egg-laying chickens turn to for compensation for laying all those extra eggs for tricky Farmer Brown?

Every morning, long before dawn, when the moon was still high up in the sky, Farmer Brown would sneak into the hatchery and switch the bright lights on. All at once the chickens would awaken and start their loud endless clucking. You could actually hear them telling each other what a bright, beautiful sunny day it was, without suspecting that it was Farmer Brown who made the dawn come up like thunder. Even the roosters thought the ruse was unfair.

WHY DO LIFE SPANS OF ANIMALS DIFFER?

The life span of each animal is determined by heredity. The rate of an animal's metabolism, how fast the heart beats, how fast the animal breathes, all have an effect upon an animal's life span. The faster the breath, heartbeat or metabolism, the shorter is the span of life. For example, the heartbeat of a mouse is 650 beats per minute and its span of life is 2 to 6 years. The heartbeat of an elephant is 35 beats per minute and his span of life is 60 years. A man's life span is about 70 years, and his heartbeat is about 70 times per minute. Only environmental changes and other conditions over a period of millions of years can effect a change in the span of life.

WHY DOESN'T A SPINNING GYROSCOPE FALL?

The gyroscope is the modern version of the ancient toy known as a top. It is a wheel mounted in a ring so that its axis is free to turn in any direction. When spun rapidly, it can keep its original plane of rotation, no matter which way the ring is turned. Thus, as long as it possesses angular momentum or the speed of rotation, it will stay in position.

But it is like a cyclist or water skier. A cyclist will fall if he slows almost to a stop, and a water skier will sink if he is not in swift motion in the water.

Today gyroscopes are very important to navigators

on ships, planes and rockets, as they maintain their course using them as an automatic pilot.

A gyroscopic compass will indicate true geographic north because the spinning wheel in the compass is not influenced by the magnetic variations of the earth, nor by the deviations from the vessel from which it is operating.

HO HUM. WHY DO WE YAWN?

If your boss catches you yawning, and thinks you're bored with your job, tell him your hypothalamus gland, a small organ located in your brain, made you do it. It controls your yawning and you had nothing to do with this involuntary reflex.

"A reflex? Then why was your mouth open?"

Tell him your mouth was open because the muscles were fatigued and relaxed and you were trying to get more oxygen into your lungs to shake off the drowsiness. By this time, if you're still yawning, I'm sure your boss is yawning too.

WHAT ARE SUNSPOTS?

The spots that appear on the sun from time to time appear to be dark only in comparison with the much brighter surface. Exactly what causes these localized flaming upheavals is not known, but they are believed to be an electrical storm. The sunspots appear in cycles of about eleven years, when they spring up in great numbers. It is during these appearances that radio and other electrical communications on our planet report severe interference. It is also believed that sunspots are the cause of the electric effects of the northern and southern lights, known as Aurora Borealis and Aurora Australis. By watching the sunspots and their daily changing positions, it was learned that the sun rotates on its axis once in about 26 days.

WHEN DOES BOILING WATER MAKE A GLASS CRACK?

An expensive piece of glass or china, made to hold a liquid, has little variation in its thickness throughout the entire body. But a cheaply manufactured glass or jar has a varied thickness throughout. We know that when heat is applied to a substance, it will expand. So when boiling water is poured into a cheap glass, the glass molecules react to the heat of the boiling water and begin to expand. Since the glass has variable thicknesses, it will crack at a point of thickness where it cannot expand as rapidly as the thinner part. Generally, glass is a poor conductor and will expand rapidly when heat is applied.

WHERE DID THE BUTTERFLY COME FROM?

If you told me that Rip Van Winkle actually awoke as a fully dressed astronaut ready for takeoff, I would say it hardly compares to the furry caterpillar emerging as a mature butterfly in its dramatic color scheme on four beautiful wings.

Starting out as an egg, laid by mother butterfly under a leaf to keep it safe, in a week or two—or maybe months for some species—a small caterpillar emerges. Soon it begins to feed voraciously on vegetation, until it grows so fat it becomes too big for its skin. Before going into the pupa stage, it sheds its skin four or five times. After the last shedding, the skin becomes hard and forms a shell called a chrysalis.

It then fastens itself under a leaf and with sticky threads of silk wraps the leaf around itself. Here it remains inactive for weeks or several months, depend-

ing on the species and the time of year. When at last it emerges, it is a gorgeous creature with all the dramatic bisymmetrical color scheme on four beautiful wings, a thing of beauty, a butterfly.

This change from egg to butterfly is one of the world's greatest mysteries and wonders of creation.

WHY WILL A NICKEL SINK IN WATER AND FLOAT IN MERCURY?

A liquid exerts an upward force on objects placed in it. It is called "buoyant force." A person weighing 150 pounds will therefore weigh less in water, due to this force. Also, the weight of the liquid displaced by the body determines whether the body will sink or float in it. A cubic inch of lead will sink in water because it weighs more than the amount of water it displaces, namely a cubic inch of water. For the same reason a nickel will sink in water but float in mercury, because the quantity of mercury displaced by the nickel weighs more than the nickel. This is Archimedes' principle, first mentioned about 240 B.C. Incidentally, he was the one who started calculus. We who are not so good in math wish that he had never brought the subject up.

DO THE FEATURES ON THE MOON'S SURFACE EVER CHANGE?

The principal cause of lunar features is now thought to be either volcanism or impacts from objects from space. This is believed to be the sole extent of lunar surface change. We know that surface water, winds, storms and erosion are so important to land features on the earth, but since these are nonexistent on the moon, the surface features of the moon hardly ever change.

HOW DOES AN OYSTER MAKE A PEARL?

It takes a foreign substance, like a grain of sand which accidentally enters the shell, to produce a pearl. In trying to reduce the irritation which the foreign grain gives to the oyster, the oyster deposits successive layers of nacreous material (mother-of-pearl) around it.

It is more the activity of the mussel family of mollusks than the common edible oyster that produces the pearls. These are salt-water, black-shelled oysters, bivalves that are also used in making beautiful buttons.

To cultivate a pearl, a tiny particle is inserted in the pearl oysters, and a wait of three to five years must take place before a pearl is produced. A worthless lump of calcium could be born even after this long wait, if the sea conditions are not just right during the waiting period. Yet, even under the best conditions, only 60% of the cultivated oysters live to yield pearls, and only 2% to 3% produce gem quality pearls.

WHY DO BLIMPS USE HELIUM INSTEAD OF HYDROGEN?

Although hydrogen is lighter than helium, it is flammable. Helium is an inert colorless gas, and though it is only 92% as buoyant as hydrogen, it is safer because it does not burn. Therefore it is safe for toy balloons and airships. Breathing mixtures are now made of helium and oxygen instead of air, because it does not form bubbles in the blood the way nitrogen does when a diver gets the bends. When the German dirigible named *Hindenburg* burst into flames on May 6, 1937, while approaching Lakehurst, New Jersey, the exploding hydrogen-filled gas bags killed thirty-six people. This brought about the change to the use of helium instead of hydrogen.

WHY DO FROZEN POP BOTTLES EXPLODE?

When water turns to ice, it expands. In a bottle of soda pop, not only is the bottle capped, which prevents expansion during the process of change, but the soda contains bubbles of carbon dioxide gas which keeps increasing in pressure while the water freezes. The combination of the two factors causes the explosion.

HOW DO BIRDS LEARN TO BUILD THEIR NESTS?

Birds don't learn to build a nest. They are born with a nest-building instinct. What materials do they use? Anything available in their environment. Teddy Roosevelt once said, "A person should do the best he can, with what he has, where he is." It seems that the

little birds knew about this long before Teddy's time. Without plans or layouts or lessons from a building school, each kind of bird builds a nest suited to its own needs. What's more, each species always builds the same type of nest its ancestors built.

Some types of material used are clay, mud, grass, strips of palm leaves, dried twigs, cotton strings, pieces of cloth and silk, stones and pebbles. Some birds build in holes of trees, in cliffs, caves or on the ground. Some build rafts from decaying plants and tie them to reeds growing in the marsh. The chimney swift uses its saliva to glue tiny dead branches together for its nest inside a chimney. But the tailor birds of India and Africa really take the prize. If your socks need darning, and you don't mind having it done with thin long strips of green grass, you can call on the little tailor birds to make repairs. With their little beaks, they build a cup-shaped nest by sewing leaves together. No one has ever reported that the stitches didn't hold.

WHAT DO WE MEAN BY STREAMLINING?

Nothing in this world can have perpetual or near-perfect motion so long as there is friction to contend with. There are various types of friction, such as ground, wind and water. For years scientists have wrestled with the problem of reducing or eliminating as much friction as possible. Since we know that it cannot be eliminated entirely, except, of course, in outer space, the less friction encountered, the faster the object will travel. So they studied the bodies of the swiftest birds and fishes, and decided that for moving bodies such as cars, trains and planes, the raindrop shape is the most efficient design to streamline a body and cut down friction. Thus, in a streamlined body resembling a raindrop, the air moves around the curved front and tapers off along the sides. This results in the least amount of drag or friction and provides the greatest speed.

Incidentally, do you remember sliding down a handrail called a banister when you were small? And did your hands and the seat of your pants get tingling hot? That was friction.

WHAT IS A "BLUE BABY"?

When our blood contains the proper amount of oxygen supplied by the lungs, it has a deep red color. When lacking in oxygen, the blood takes on a bluish color. Thus, when a baby's appearance is bluish in color, it is due to the lack of sufficient oxygen in its blood. A malformation of the walls between the chambers of the heart causes the blood from veins and arteries to mix. This reduces the effectiveness of the baby's lungs in their job of oxygenating the blood. Since the blood lacks enough oxygen, the baby appears blue. When the baby is old enough to permit surgery, the doctors sew together the gap in the walls of the heart and the condition is corrected.

WHAT MAKES PLANTS LEAN TOWARDS THE SUN?

Have you ever watched a plant try to reach for sunlight? If you put a plant in a dark shoebox that has a small hole in it to let in a bit of light, you will see how it will bend and twist to get to the light. If you could refer to the long stem of the plant as its neck, you might even say that the plant will break its neck to reach the light. There is a certain chemical called auxin which makes the stem grow faster on the dark side. Since now there are more cells on the dark side, the plant naturally will bend towards the light. This is called phototropism.

HOW STRONG ARE THE GROWING ROOTS OF TREES?

An example of nature's power is the chunks of mountain that break away and come tumbling down each winter. In the fall and winter, when rain gets into spaces between the rocks, the freezing cold makes the water expand when it turns to ice. Since the ice has to take up more space than the rainwater did, because it grew larger as it froze, it caused the chunks of mountain to break away.

A similar action takes place underground that shows the power of nature, in the strength of growing roots. The power that lies hidden in the roots of trees can easily be seen when you walk in the streets of a city. If a tree is growing through the sidewalk, stop and take a good look around its trunk near the ground. If the sidewalk is made of concrete, you will surely see some cracks in it. This will give you some idea of what the growing roots of a tree can do. And that concrete may be as thick as six inches.

The roots of a growing Ficus tree have been known to completely destroy the plumbing of a building, and, in another case, to have lifted a house from its foundation.

This writer has actually seen peagrass so strong in its upward push to reach sunlight that it penetrated through a one-half-inch-thick layer of asphalt pavement.

WHY CAN A HEADLESS ANT SURVIVE?

Unlike the vertebrate animals, insects have no spinal cord to form a central nervous system. Instead, they have little knots of nerves called ganglia, connected to paired nerve cords at the underside of their bodies. Each ganglion controls a certain phase of activity, and is often capable of functioning by itself. It is like the big valves on a water dam that can be shut off individually. You can shut off each valve for a certain section, but the dam will function where the valves remain open.

To prove this, it was shown that ants with their heads cut off were able to continue to walk and live on for some time. One ant lived for over a month in the laboratory where he was beheaded. He just didn't use his head.

WHY IS PORCELAIN MORE EXPENSIVE THAN OTHER POTTERY?

The type of chinaware known as porcelain is a fine translucent earthenware of superior whiteness and hardness. It differs from other pottery because it is made out of special material. To make it, manufacturers use a special china clay called kaolin, and china stone called petuntse. When shaped and decorated, it is fired in a kiln from 1,300° C. to 1,500° C. Hot liquids suddenly poured into china will not make it crack, because it is a good conductor of heat and does not expand as rapidly as ordinary glass does.

WHY DO CAMPHOR BALLS DISAPPEAR WHEN EXPOSED?

The camphor molecules keep flying off into space, because, as in every other substance, they are constantly in motion. It may take a few months or more, but the camphor ball, which has about a three-quarter inch diameter, will continue to shrink until it completely disappears. It is a form of evaporation like water molecules jumping out of a birdbath. When you come to think of it, it's a good thing our dead skin cells regenerate, or else we'd end up like the camphor ball.

HOW DO GRASSHOPPERS AND CRICKETS MAKE SO MUCH NOISE?

Crickets have scrapers and files on the basal part of their wings. When they rub these together, the entire wing membrane vibrates. The short-horned grasshopper vibrates its wing membranes by means of a row of fine projections on the inner surface of the hind legs which it rubs together.

ARE THERE BULLIES AMONG ANIMAL GROUPS?

It seems that social hierarchies do develop among certain animal groups. For example, there is such a thing as a "pecking order" in a flock of chickens. In a study of one flock it has been discovered that some chickens may peck at others without retaliation. There is one bird in the flock who is free to peck at any other bird and not be pecked in return. Another bird may freely peck at any other bird in the flock except this first one, and so on.

WHY DO WE SNEEZE?

A sneeze is an involuntary sudden action to exhale breath from the nose and mouth. Many things may cause a single or series of sneezes. An irritation of the nasal mucous membrane by dust, pollen or tobacco smoke can bring it about. Violent fits of sneezing can be caused by hayfever, allergies, asthma, or whooping cough, but the common head cold is the greatest offender. What one should remember most about sneezing is that thousands of bacteria get a free ride in the atmosphere on tiny droplets for as much as 20 feet. So remember this advice: If you don't cover up to protect your neighbor, it's nothing to be sneezed at.

HOW TRICKY CAN A SPIDER GET?

The trap-door spider lives up to its name. It digs a burrow in the ground and covers the entrance with a lid or trap door on a hinge. Several kinds of these spiders live in warm or temperate regions. These large hairy creatures of the tarantula family line their burrows with silk. The trap door itself is made of layers of silk and mud and fits so perfectly it keeps the water out. One spider in southwest United States has a tunnel eight to ten inches long and one inch wide, which it usually builds on high clay soil where it will stay dry.

Trap-door spiders work at night and take about 16 hours to make a burrow. When no one comes near their trap door, they hunt at night and eat ants and other insects, caterpillars and earthworms. Usually they wait discreetly under the trap door for their prey. When they feel the footsteps of an insect, they pop open the door, catch the victim and drag it in. Sometimes they conceal the entrance with a layer of earth or gravel. In

Virginia, the trap-door spider covers its door with living moss.

Female trap-door spiders may live over twenty years. They can lay over 300 eggs at one time in a silken sac from which the young emerge in about two months. The young will nest with the mother for about eight months, then leave to build burrows of their own.

HOW WELL CAN WE SEE IN THE DARK?

Leave it to mankind to invent a device to destroy itself, not only during daylight, but also in the darkness of night. An invention called a snooperscope is an electric device which makes it possible to see objects in the dark. It throws out an infrared light beam, and an electrically operated telescope sight makes an invisible object form a clear image in green on the snooperscope. Any moving object or form can be seen in the dark.

Another device, invented for use in the Vietnam war, electronically amplifies light to produce a far brighter image. This allows one to see in the dark, without a beam of any light other than that reflected from the object by moonlight.

HOW DO BIRDS KNOW WHEN TO MIGRATE?

Many species of birds migrate twice a year. In the fall they fly to a warmer climate and in the spring they return to a cooler climate along well established air routes. Much evidence supports the theory that the amount of sex hormones in the bird's blood is the stimulus for migration. The concentration of sex hormones in at least some birds varies with the length of day. Since the days begin to shorten in the fall, that's when they begin to migrate.

WHAT FISH HAS A BUILT-IN TIME CLOCK?

The grunion is a small, edible six-inch fish with a solid silver streak on either side of its body. It swims in the waters of the Pacific Ocean from the coast of San Francisco to lower California.

What is most mysterious and baffling to scientists is the timing of its spawning habit with clockwork precision. In the first place, the fish appear from March to June when the highest tides of the year roll around. Not only do they wait for the highest tides, but with an instinct possessed only by them, they roll onto the beaches on the highest wave. They seem to know that the correct spawning time is on nights of the highest tide in the new-moon period, or two weeks later during the full-moon period.

Now comes the phenomenal act. No sooner have

they landed than the females dig themselves a hole in the soft sand with their tails in rapid motion until their heads are barely visible. As soon as the eggs are laid they crawl out of the hole, once again using the powerful force of their fins and tail. The male grunion, who was anxiously waiting nearby, immediately backs into the same hole to deposit his sperm and fertilize the eggs.

The wonder of it all is that this rapid activity must be precisely completed so that the next high wave that appears will carry them out to sea again. If the high wave does not carry them out to sea, they will die on the sandy beach.

ARE ANY LIVING THINGS IMMORTAL?

The late Prof. L.L. Woodruff of Yale University followed the detailed history of a few individuals of the slipper animalcule paramecium through 11,000 counted generations. One cell divided to become two animalcules, the two to become four, and so on. Potentially, each cell seemed to be immortal, and Prof. Woodruff had no way to know how many years or thousands of years this same process of asexual duplication would go on without any slipper animalcule dying of old age.

Another example of what seems to be an immortal life is that of the hydra. The individual cells of the hydra may never be more than a month old, since they arise in the collar-shaped ring below the tentacles and migrate slowly to the regions where they are cast out of the body. The hydra, as an interconnected population of cells, seems able to maintain its individuality indefinitely because it constantly renews itself.

The coral polyp, which greatly resembles the hydra, should be included in this group.

WHAT BIRD IS THE MIGRATION CHAMP OF THE WORLD?

The Arctic tern is considered the champion because twice a year it makes a flight of 11,000 miles. This bird, only seventeen inches long, with a blue-gray back and white feathers, breeds on the seacoasts from New England north to Greenland, and the northernmost islands of the Arctic Ocean. In late August, the young terns and their parents start their long journey to the shores of Antarctica and the nearby islands. After a stay of several months, they fly north again, arriving in Arctic regions in mid-June. The real mystery to this 22,000-mile trip is why the Arctic terns do it at all.

WHAT MAKES A CAT PURR?

A cat is born with two sets of vocal cords. One set, in the voice box, makes the familiar "meow" sounds. The other set, the false vocal cords, are vibrated on inhaling and exhaling, producing an involuntary continuous purring sound, when the cat is in a contented mood.

WHAT IS DRY ICE?

Dry ice is carbon dioxide gas compressed until it is liquefied, and then frozen. It is used as a refrigerant (freezing agent) because it changes back to a gas without becoming liquid. It remains in a solid state for a longer time and at a colder temperature than does ordinary ice. It gets its name "dry ice" because it doesn't turn back into a liquid.

WHAT DO WE MEAN BY THE RH FACTOR?

The Rh factor is a substance found on the surface of red blood cells which causes them to stick together. It is called Rh factor because it was originally discovered in rhesus monkey blood. If you have it your blood is considered Rh positive, if you don't it is considered Rh negative. Like the A, B, and O factors, it is essential to know whether a person's Rh factor is positive or negative when matching blood types.

If an Rh-positive man marries and Rh-negative woman, the first child born is usually safe. But the second child will be born with a condition that destroys the red blood cells, leaving the baby severely anemic with accumulated toxic substances in its body.

Even though the mother was Rh-negative, she will be exposed to the Rh-positive fetal blood of the baby which was inherited from the father and now goes into her system. This makes it impossible for her to have future Rh-positive babies, because she begins to produce antibodies in her blood that attack and destroy the baby's red blood cells, thus causing the baby's death. About 12% of American marriages bring this type of child into the world. But now medicine has a remedy.

Today a form of blood extract called Rh immune globulin acts as a vaccine to curtail the Rh-negative woman's production of antibodies and greatly reduces the risk to future Rh-positive children. It must be given to every Rh-negative woman within 72 hours after her first delivery to prevent her from becoming totally immune to future shots.

DOES THE SEA GROW CUCUMBERS?

A certain marine animal, related to the starfish family, has a cylinder-shaped body and resembles the garden cucumber. That's how it got its name. In the ocean where there are hundreds of species, they grow two to three feet long in tropical waters, and a few inches to a foot long in the temperate regions.

At one end of the cucumber is the mouth opening surrounded by ten branching tentacles that catch the food. There are five double rows of tube feet which can be used for walking when each little tube is extended. Those caught in the East Indies are dried and sent to the food markets of China. In other countries they are called Trepanz and Beche-de-Mer.

These sea cucumbers can detach and throw out their intestines when another animal attacks them. This hampers the movements of the attacker while the sea cucumber makes its getaway. The creature has nothing to lose but time, because eventually it generates a new set of intestinal organs.

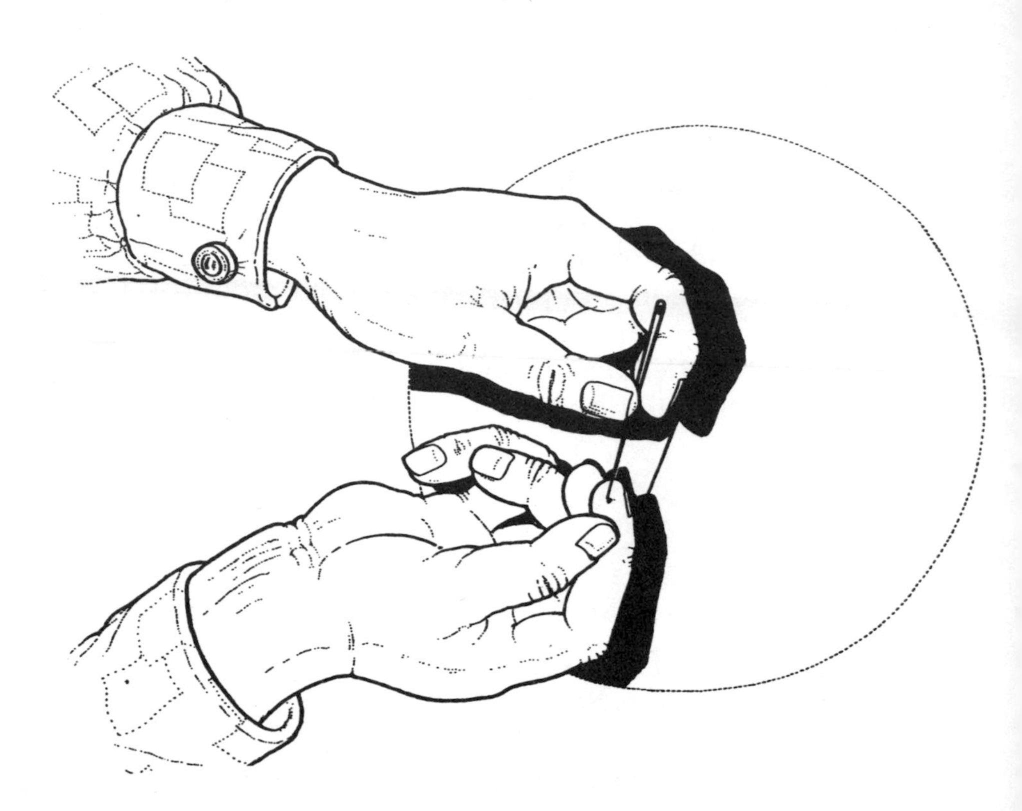

WHAT TYPE OF BLOOD HAVE YOU?

All human blood is not alike, as was discovered around 1930 by Dr. Karl Landsteiner, a Nobel Prize winner. He classified blood into four different types, depending upon whether or not the red blood cells clumped together in a transfusion. In transfusion it is important to know what type the patient has, for if the red cells clump, a violent lethal reaction could set in. This system is now used internationally and the four blood groups are O, A, B, and AB.

Type O. The blood of a person with this type can be freely given to any of the other groups without fear of clumping. But the owner of this type can only receive this type and nothing else.

Type AB. This blood if given to the other groups would cause clumping. But a person with this type can receive blood from all the other types without any harm.

Type A. This type will clump with B.

Type B. Will clump with A.

There are two other factors in blood, called M and N. By a study of the blood of the supposed father of a child, it is possible on the basis of these two groups (if certain other factors are present) to determine that the man could not have been the father.

HOW DOES A THERMOSTAT WORK?

All people do not react to the heat or cold in the same way. What is too hot for you may be just right for another. Different metals in a way are like human beings. The heat expands different metals at different temperatures, and the cold contracts different metals at different temperatures. This gave a scientist an idea to make a tool that could start or stop a motor or an electric switch. Because it could control the heat of a room he called it a thermostat.

A typical thermostat is a piece of brass and a piece of steel welded together in the shape of a question mark. As the metals expand or contract with the heat or cold, the tail end of the thermostat moves one way or the other. This movement from left to right can start an air conditioner or a heater, depending on how you set the room temperature on the thermostat.

HOW DEEP CAN DIVERS GO?

A diver connected to the surface with communications and air tubes may go down to a depth of about 500 feet or slightly more. This is the emergency limit for Navy divers. A diver could spend about three or four hours working at a 200-foot level, but his return to surface would have to be very slow.

Scuba divers of course have more mobility, but even with special diving apparatus, they are limited to less than 200 feet. Westinghouse Underseas Division has recently created a new "closed circuit scuba" outfit which, with its safe breathing mixture, permits diving to a depth of 1,000 feet or more.

Jacques Cousteau and Edwin A. Link have tested underwater structures that make it possible for divers to live and work under water for weeks at a time.

WHAT IS AN ANTIBIOTIC AND HOW DOES IT WORK?

When you are sick and certain germs are causing your illness, these germs are called pathogens. In order to try to make you better, the doctor will give you an injection with a medicine that will fight these pathogens and kill them. This medication is called an antibiotic. It is like a soldier who fights these germs to save your life.

Penicillin is an antibiotic because when it is sent into your bloodstream it kills the germs that are making you sick. Sir Alexander Fleming discovered this antibiotic in 1928, and for this he won a Nobel Prize in 1945.

WHAT ARE ALGAE AND WHY ARE THEY IMPORTANT?

This one-celled plant, of which there are 2,000 species, contains chlorophyll but no true root system or leaves. It is found in water and in damp places on shore, on stones, rocks and trees. Some live on animals such as turtles, waterfleas and sloths, and some even live in the intestines of animals. Some species float as masses in lakes and oceans and are known as plankton.

They may take any conceivable shape like kelp or devil's apron, and reach 30 feet or more in length and two or three feet in width. Although they are plants, some move by twisting or gliding. Many algae fall into the four color groups of blue-green, green, brown and red. The fifth group of these one-celled plants are called diatoms.

The green algae are food for fishes and the brown algae are usually in salt water and are called seaweeds. Kelp is one of these. The brown algae are used as fertilizer, and from the ash that is left after burning it, we get iodine and potassium. In the Orient, not only are the brown algae used as fertilizer, but also as food for both humans and livestock.

IF THE SUN BLACKS OUT, HOW SOON WILL THE EARTH?

Since the sun at its furthest point from earth is about 94 million miles away, and since light travels at a speed of about 186,000 miles per second, it would take about 8 1/3 minutes for our world to reach total darkness. But it won't happen for another billion years.

WHY DOES AN UPRIGHT STICK LOOK BROKEN IN WATER?

Did you ever notice how a stick looks broken or bent when held upright in a glass of water? Because water is denser than air, light rays travel faster in air than in water. Therefore there is a change in the path of the light ray. This is called refraction of light. The light waves bend while going through a semisolid like water. That is why you get the illusion of a bent or broken stick.

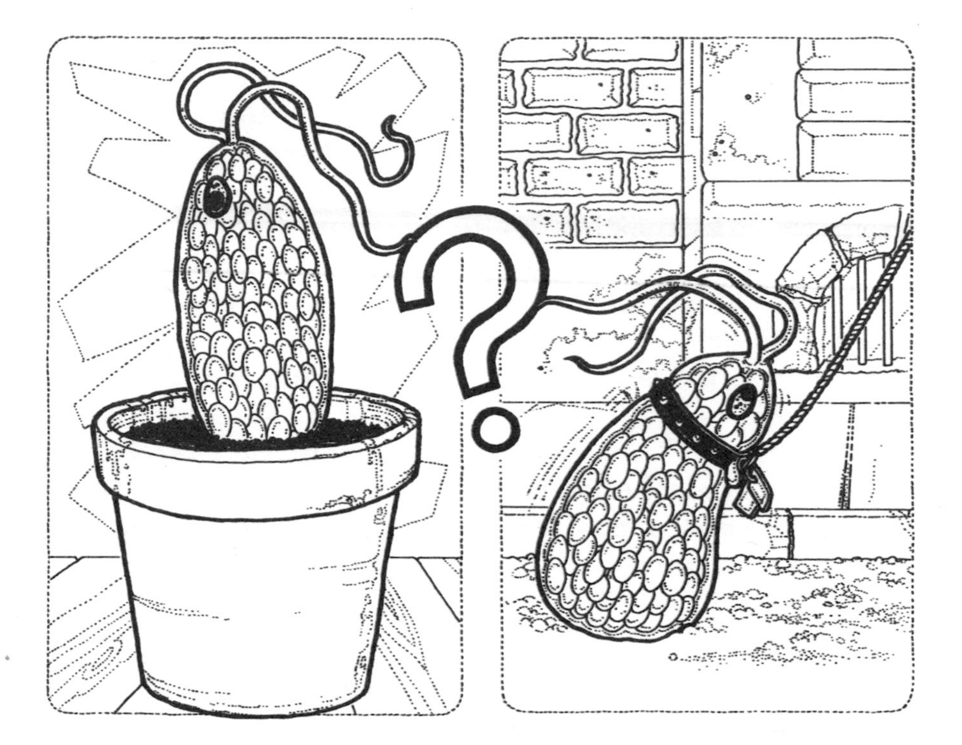

ARE YOU SURE YOU KNOW A PLANT FROM AN ANIMAL?

According to the scientific classification of plants and animals, most plant cells contain green chlorophyll and animal cells do not. Well, along comes the one-celled something-or-other called Euglena that puzzles the scientists because it has the characteristics of an animal, yet contains chlorophyll cells, which only plants have. Like all animal cells, it does not have a cellulose wall. But it has mobility, a trait all animals possess at some time in their lives, and it moves about in the water by means of two flagella fused together. It also has an eyespot. When kept in the dark, it loses its chlorophyll like plants do. But when restored to light, it develops chlorophyll again, just like plants do. So, here we are again. What is it? Plant or animal?

WHAT CAUSES A TIDAL WAVE?

Tidal waves should not be associated with true tides that are governed by the sun and moon and have regular periods of appearances. Tidal waves are caused by undersea earthquakes called seaquakes or by hurricanes out at sea. In open seas the waves travel between 400 and 500 miles per hour. Experience has shown that earthquakes in the Aleutian Islands cause tidal waves in Hawaii. Now seismographs can tell, hours before, when a tidal wave will reach Hawaii, by the record of an earthquake in the Aleutians.

Tidal waves may be only two or three feet high on the open ocean, but when they approach land they reach as much as 50 feet high, such as the one that wrecked Lisbon, Portugal, in 1755.

HOW DO YOU EXPLAIN DREAMS?

The answer depends upon whom you ask. An American Indian would tell you it was a message from the gods. If you were superstitious and dreamed of a house with seven gables, you would rush to your bookie and place two dollars on horse number seven in the seventh race. If you asked Sigmund Freud, he would say it was a way of achieving unfulfilled hopes and desires. Like Thurber's character Walter Mitty, he could have instant strength, riches and power just by lowering his lids and dozing off. If you ask dream researchers they will tell you that dreams express your personality. If you're heroic, you'll be the hero in your dreams. If you're timid, you'll wake up just before the fight. You may have four to six dreams during the night at 90-minute intervals, each lasting about thirty minutes. The first dream, which comes about an hour after dozing off, will only last a few minutes. But don't lie in an uncomfortable position, because that leads to dreams of falling and struggling. Pleasant dreams.

WHAT SUPPLIES MOST OF THE OXYGEN IN OUR AIR?

Diatoms are microscopic algae consisting of one cell. Several thousand species live in fresh and salt water seas. They propel themselves and consist of two halves called valves. The smooth glassy shell is usually silica and is insoluble in water. That is why the sea and ocean floors contain abundant deposits of the valves of long deceased diatoms.

Diatoms are important as food for small sea animals which in turn are food for larger fish. In addition, one of the most important functions of the diatom is to give off oxygen as a waste product during photosynthesis. Because of their tremendous abundance, diatoms contribute the greatest amount of oxygen to our atmosphere.

CAN FRESH WATER BE MADE CHEAPLY FROM SALT WATER?

At this point the answer is no. Desalination is not yet an efficient way of getting water either for drinking or irrigation. Commercially it still does not pay, except in a few water-poor areas where desalinated water is economically better than no water at all.

The increase in the use of water in the United States is at an unbelievable rate of 1.5 million gallons an hour. In many places water shortages were so critical that in 1952 Congress passed a Saline Water Act to establish the Office of Saline Water, in order to develop a low-cost commercial way to increase the supply of drinkable water. It is entirely possible that from the original cost of $4 per 1,000 gallons, the price may eventually come down to 20¢ per 1,000 gallons.

In 1972 a nuclear-powered desalination plant started operation in southern California. Its capacity reaches 150 million gallons per day, and it gives 1,800 megowatts of electricity. It can produce more fresh water than all the desalination plants operating in the world in 1966.

Although the oldest method, known as solar distillation, is not economically feasible, there are areas in the world where small quantities of water are needed but energy sources and technical competence are not available. In these areas the simple solar method may still be a partial answer.

HOW DO HEART VALVES WORK?

Separating the auricles from the ventricles in the heart are small flaps of tissue called valves. These allow the blood to flow only in one and the proper direction. Should the blood try to go the wrong way, the valve closes and won't permit this to happen. Although there are three types of valves in the heart—semilunar, tricuspid, and bicuspid—their function is the same: to keep the blood flowing in one direction. The veins also have valves which prevent the blood from flowing backward when the blood is on its way to the heart.

Certain diseases, like rheumatic fever, can cause the heart valves to lose their one-way function by weakening the hinge point to the extent that it allows some blood to flow back. This places a greater burden on the heart.

WHAT IS A MAGNET?

A magnet is an iron bar or horseshoe-shaped steel rod, in which the molecules of iron are all turned with their magnetic charges and lined up in one direction like soldiers. Normally these molecules are scattered and irregular in arrangement. It is amazing to note that when an ordinary iron bar, whose molecules were in total disarray, was magnetized in an experiment, the sound of the molecules, turning to be arranged in orderly and soldierly fashion, was heard and broadcast over a radio.

These are some of the features of a magnet. It has a north and south pole, each of which points to the opposite north and south magnetic poles of the earth. It has the power to attract iron, steel, cobalt, and nickel. If you should accidentally drop the magnet on solid ground, thus jarring and disarranging the molecules, that will be the end of your magnet.

HOW DOES A SKUNK DEFEND ITSELF?

It may not be what the English would call "cricket," but compared to the defense weapons of other animals this weapon can certainly be extremely effective.

The skunk's peculiar means of defense is a pair of glands with nipplelike openings under the tail. These

glands contain a fluid with a strong, sickening odor. When frightened or attacked, the skunk can squirt this fluid with considerable force and accuracy. This keeps enemies who suffered a previous encounter at a safe distance. A new enemy will have something to remember for a long time, and it won't be romantic. The odor will linger for many days.

However, this attractive little animal, about the size of a large cat, has been made into a domesticated friend by many animal lovers. But to insure a binding friendship they first bind the nipples of the glands with gut to prevent an unpleasant accident. Eventually the gland dies from lack of blood supply.

WHO ARE THE SOLITARY BEES?

Some people love city life among urban multitudes and some choose the solitary life of the rural farmer. Well, even the bees like a choice. There are solitary bees and social bees. The social honeybees live in hives inhabited by thousands, each one doing his share to maintain the hive. The queen lays the eggs, the drone mates with the queen, and the workers gather food, care for the young and keep the hive in order. But solitary bees live in small colonies wherein each family has its special apartment. There are no queens in these colonies.

The black carpenter bees bore a one-inch hole in wood to build their nest. It has partitioned walls with one egg in each partition, and when completed the nest contains from three to seven cells.

The mason bee builds cells for eggs out of moist clay mixed with small stones and bits of sticks and leaves. She will burrow a tunnel in a piece of wood and apply the clay. When the inside is smooth, she places one egg and a store of honey and pollen in each cell. Then she covers the cell with more clay. Later the young force their way out.

The leaf cutter bee cuts pieces of leaves and glues them together to form a small thimble. She then lays an egg in each thimble and closes it with another piece of leaf. The nest may contain from three to twelve thimbles.

The cuckoo bees are the parasites and the cheats, because they make their homes in the nests of bumblebees. They will do no work to maintain the nest, but will steal the food and lay their eggs in the nests of other bees.

The miner bees dig tunnels in sandy banks underground. Each female then digs a separate compartment alongside the main tunnel to lay eggs and raise her brood. A nest may contain hundreds of compartments and thousands of bees.

HOW WAS PENICILLIN DISCOVERED?

Sir Alexander Fleming, a British bacteriologist, is the person to whom all mankind owes a great debt of gratitude. An accident that occurred in one of his experiments might have been overlooked by any number of other scientists. But his inquisitive mind led him to further inquiry and research that eventually paid off seventeen years later with a Nobel Prize for his contribution to the field of medicine.

In 1928, working in his laboratory with a culture plate of bacteria, he accidentally dropped some green mold on the plate near the culture. This mold was *Penicillium notatum.* A short time later, Sir Alexander noticed that near the spot where the mold fell, all the bacteria were dead. Seeing this accident as a potential benefit to mankind, he and several associates began experiments with the mold that lasted for many years.

Today, all the world knows of the success of penicillin in treating hopeless cases of disease, injuries and burns. It is used in the treatment of venereal disease, diphtheria, meningitis, tetanus, middle ear infection, eye infection, blood poisoning, heart infection and countless other ailments.

Though billions of units of penicillin are manufactured each month, and though many are semisynthetic, basically they all come from the mold *Penicillium notatum.*

WHAT IS AN ARTESIAN WELL?

At another time, I might be inclined to tell you how to make a Venetian blind. Right now, I prefer to tell you how to make an Artesian well.

An Artesian well is formed by boring into a layer of porous rock (sandstone) or loose sand, from which pure water flows under its own pressure. This layer lies between two layers of harder rock or clay. Such wells can be found all along the eastern coast of the United States and in Europe. Pittsburgh, Pennsylvania, has one that is 4,625 feet deep, and one located in Leipzig, Germany, is 5,735 feet deep.

The Artesian well was named after the province of Artois, in France, where the first well of this type was drilled in the twelfth century.

WHAT HAPPENS WHEN WE BLUSH?

In certain emotional situations, our face and neck turn red and warm. This comes about because certain nerves called vasodilators are stimulated and make the tiny blood vessels in the face and neck expand. This allows more blood to flow through them, thus giving them a red appearance. Getting caught in a lie could bring it about.

HOW DOES A FARMER FIGHT EROSION?

The following list shows some of the things he does.

1. He plants grass in gullies to act as a filler.
2. He uses fertilizer to replenish lost minerals.
3. He makes plateaus and terraces on hilly land.
4. He plants in alternate years.
5. He plants in alternate rows, called strip farming.
6. He plants new trees and replaces those that die.
7. He alternates yearly crops with alfalfa or other nitrogen-giving plants.
8. He does contour plowing around a hill.
9. He uses irrigation systems on his farm.

HOW IMPORTANT ARE MINERALS TO OUR BODIES?

The phosphorus in bananas, dairy products, and green and yellow vegetables is good for our nerves and muscles. The iron in raisins, liver, whole grain cereals, egg yolk, beans and leafy vegetables is good for our supply of red blood cells. The calcium in milk and other dairy products and vegetables helps build our bones and teeth. The iodine we get from fish and iodized salt aids in the function of the thyroid gland. The sodium in salt helps build new tissue, and the chlorine derived from the same source makes the hydrochloric acid so essential to our digestion.

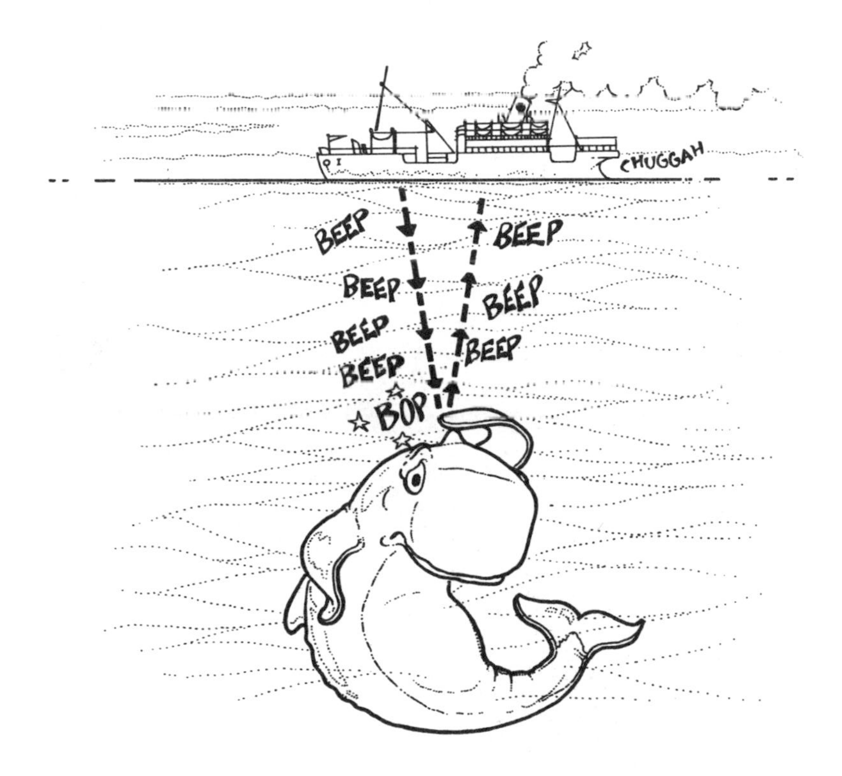

WHAT IS SONAR AND HOW IS IT USED?

Sonar is a device to locate objects under water by the reflection or the echo of sound waves. To send out sound signals, short sound waves from a high frequency vibrator are used. Thus, if we know that sound travels through water at a velocity of 4,760 feet per second at 15° Celsius, and a signal is sent out to an object under water to an unknown distance below, the measured time it takes for the echo to return, divided in half, will give the depth of the object below the surface.

IS MERCURY DEFYING GRAVITY IN A THERMOMETER?

When body heat is applied to a thermometer, it is sufficient to expand the atoms of mercury and push the mercury up. It comes from the simple principle that heat applied to substances expands them, and is the same basis upon which the thermostat operates. The pressure of expansion due to the heat applied is sufficient to overcome the force of gravity which tends to pull it down. Under your tongue the thermometer gives a more correct reading and helps keep your mouth shut for a little while.

HOW HIGH CAN A WAVE GET?

In 1933 while going from Manila to San Diego, an American tanker, the USS *Ramapo,* reported a wave 112 feet high, brought on by a wind of 60–68 knots (66 to 74 miles per hour). This was the highest wave ever measured with any degree of accuracy. While 25-foot waves are rare, and most ocean waves are less than 12 feet high, severe storms can develop waves as high as 50 feet.

HOW COME WE ALWAYS SEE THE SAME SIDE OF THE MOON?

The moon, like the earth, has no light of its own, and the only reason we see the moon at all is because of the light reflected from the sun. But the reason we see only one and the same side of the moon is because the moon not only rotates once every 28 days on its imaginary axis, but also revolves around the earth once every 28 days. Therefore it always faces us with the same side.

WHAT IS AMBERGRIS?

Ambergris is a waxlike substance, white, gray, brown or black in color, found in the intestines of sperm whales. Spewed out as waste, it is found in lumps on the ocean or cast on the seashore. The largest piece ever obtained weighed 1,003 pounds. When taken from a dead whale, its odor is most disagreeable. But after exposure, it acquires a sweet earthy smell, making it worth its weight in gold to the perfume industry. A pinch of it added to perfumes makes the aroma last much longer. So it you're beachcombing, keep an eye out for ambergris.

HOW DOES EXPOSED FILM BECOME A PICTURE?

The film that you place in your camera has a coating of billions of tiny grains of silver salts or halides. When the shutter is opened for a fraction of a second to allow light to strike the film, a chemical reaction occurs that changes the grains on the film to varying degrees of black depending on the object being pictured. The film is then sent to a laboratory where it is bathed in certain chemicals such as alum, acetic acid, sodium sulphite and hypo. This makes the chemical change permanent. The hypo helps remove the grains of silver halide that were not affected by the light. In the next step the film is washed and hung up to dry. It is now a permanent negative from which many pictures can be developed. The light parts in the negative will come out dark on the picture and the dark parts will be light. All the other intermediary shades help complete the picture. In a color film various chemicals are put on the film, each reacting differently when the light strikes it, to bring about the special color effect.

WHAT IS A CHEMICAL INDICATOR?

In chemistry, a chemical indicator is a substance that detects the presence of acids or bases in a substance. A piece of red or blue litmus paper is used. To detect an acid a piece of blue litmus paper is dipped into the liquid. If the blue paper turns red, the substance is an acid. To detect a base, you dip a piece of red litmus into the liquid. If the red paper turns blue, then the liquid is a base. The litmus papers, or chemical indicators, are made from lichen (liken), a plant that grows on rocks and trees. The litmus turns from blue to red because of a change of position of the atoms in one of the chemicals of litmus.

IF WATER HAS OXYGEN IN IT, WHY DO PEOPLE DROWN?

This seems like a foolish question and would not appear in this book were it not for the fact that every first-year chemistry class has at least one student who asks it. The answer of course is that the oxygen in the water is not in a free state to serve you as it is in the atmosphere. The oxygen atom is so tightly tied up with two hydrogen atoms, so chemically bound together in the water molecule, that it can only be separated by a complicated process called electrolysis. This process involves running an electric current through the water to separate the two hydrogen atoms from the oxygen atom. Only then could you make use of the oxygen, now in a free state.

HOW DOES AN ELECTRICAL TRANSFORMER WORK?

Transformers are electrical devices that increase or decrease the electric flow or voltage. High up on the poles that carry the electric wires, you will see sitting here and there a barrellike contraption which contains the inside coils wound on coils of iron, to reduce or increase the voltage coming from the powerhouse to be used in homes. One set of coils is called the primary, one the secondary, but neither one touches the other. Where the voltage is to be stepped down, as in the case of the voltage going into a household, more turns of coil are put on the primary coil and fewer on the secondary. Where a step-up is desired, the primary has fewer coils and the secondary more.

WHAT IS A GEYSER?

A geyser is a spring from which water and steam shoot up at intervals from an opening in the ground. It is found only where conditions are just right. Water is collected in a deep crack underground and is heated by a layer of hot rocks at the lower part of the crevice. The pressure of the water above prevents the water trapped in the lower portion of the crevice from boiling readily. Finally the trapped water does boil, blowing the water above it through the opening of the geyser.

"Old Faithful" in Yellowstone National Park is probably the world's most famous geyser. On the average, it erupts once every sixty-five minutes and lasts for about six seconds. The height of the eruptions is from 120 to 150 feet.

HOW DOES A PREGNANT ALCOHOLIC AFFECT HER FUTURE CHILD?

This item is not intended to include pregnant women who drink socially. But of an estimated two million women, half of whom are of childbearing age, doctors in Seattle have recently connected a pattern of serious birth defects among children to women who are chronic alcoholics. In a study of eight children born to such women alcoholics, a malformation syndrome was described in the journal *Lancet.* There was a significant growth deficiency, lagging intellectual and motor development, small head size, heart defects and subtle facial and limb abnormalities. All had subnormal intelligence, with IQs ranging from less than 50 to 83, and most were below average for their age in the performance of motor tasks. Weighing about half the weight of an average baby at birth, and about 20% shorter, they continued to be retarded in growth after birth. Even those cared for in a hospital or in a foster home were affected this way.

HOW DO TASTE BUDS WORK?

There are four different tastebuds located in four different places on the tongue to help us distinguish sweet, sour, bitter and salty. Exactly how they do this is not known. A substance can only be tasted if dissolved in solution. Taste buds are clusters of banana-shaped cells sunk into pits in the surface of the papillae. Each taste-sensitive cell has a hairlike process at its exposed tip, and is connected to a nerve ending at the opposite end. Some of the sensations commonly assigned to the sense of taste are in reality examples of the sense of smell. Many spices have relatively little taste, but affect the sense of smell powerfully.

WHAT SETS OFF OUR CHANGE TO MATURITY?

Most mammals have different waiting periods for maturity to arrive. Monkeys wait three years, lions two years, bears thirty months, dogs seven months, cats ten months, and man thirteen years. In all these examples, the period remains almost fixed. What interests scientists is the thing that sets off the change from immaturity to maturity. They believe the process starts in the nervous system, with the hypothalamus gland playing the greatest role. It acts as a biological timer, by turning on the production of hormones that starts sexual maturity.

ARE THERE ANY SEAS WITHOUT TIDES?

We know that the sun and the moon affect our tides, but to some extent tide ranges are affected also by the slope and dimensions of the coastline and the sea floor. In some restricted water areas like bays and channels, tides may reach fifty feet and currents as much as ten knots (eleven miles) an hour. Although some places like the Bay of Fundy move more than 100 billion tons of water a day, there are areas in the world that are almost tideless, such as the Mediterranean Sea, the Baltic Sea, the Adriatic Sea and the Gulf of Mexico.

WHY DOES PAPER BURN FASTER THAN COAL?

Every substance has a kindling temperature, a point at which applied heat will make it burst into flame. The more solid the molecules of the substance are, the higher the kindling temperature. Paper has a low kindling temperature and catches fire easily, but coal has a high kindling temperature and needs a great amount of heat before it reaches its kindling point to start to burn. Steel, which has a much higher kindling point than coal, will first start to melt at 1,375 degrees Centigrade. The reason a match lights so quickly with just one rub is that it is tipped with phosphorus, a very low-kindling-point substance. The slight friction of rubbing is sufficient to set the head of the match ablaze. And so a person with low kindling temperature is called a hothead.

WHAT CAN A SPEECHLESS SKELETON TELL US?

The chemicals used by a person, or any severe injury received during the person's lifetime, may give clues to his medical history. The shape and size of the bones may tell the sex, shape and size of the body. One physical anthropologist determined the sex of a murder victim by the size of the ridges above the eyes and at the back of the skull. It is known that at age 40, about three-fourths of the skull bones are fused. With that as a scale, a fair age guess can be made. An age guess can also be made from the fact that the calcium content of a person decreases proportionately with the years. In many cases the skull can also tell the race it belonged to.

WHAT IS THE GREATEST MULTIPLE BIRTH ON RECORD?

Triplets, quadruplets, and quintuplets are all recorded in the annals of medical journals, but as you can readily understand, the greater the number of babies born in a multiple birth, the rarer is the occurrence.

There is one confirmed report of live-born octuplets. There were four boys and four girls totaling nine pounds and ten ounces in weight. They were born to a 21-year-old woman in Mexico City on March 10, 1967. Unfortunately, they all died within fourteen hours after birth.

WHAT IS A CATALYST?

Photosynthesis is the process of food manufacture by plants, using water, carbon dioxide from the air, dissolved minerals from the ground, sunshine and chlorophyll. However, in the process, chlorophyll does not mix with the other substances either chemically or physically, and is therefore one of our best examples of a catalyst.

The chlorophyll neither gains nor loses any part of itself, since it does not actually take part in the process. It simply aids and brings about the process, which is about all to expect of a catalyst. But although it is only a catalyst without gain or loss, photosynthesis could not take place without it.

WHY DO SOME ALGAE PUZZLE SCIENTISTS?

Whenever the process of photosynthesis (the food-making process of plants) is involved, sunlight or some source of light is a necessary ingredient. Yet some algae have been found in the dark depths of the ocean that photosynthesize without light. This fact amazes the scientists who have specialized in the study of photosynthesis and have always considered light as one of the most essential factors in photosynthesis.

HOW DO YOU EXPLAIN THE BIRTH OF TWINS?

The two types of normal twins are: fraternal and identical. Fraternal twins are not rare, but identical twins are. Another type of twin, abnormal in nature, is the Siamese. Siamese twins are abnormal because they are born even before they have completely separated during gestation.

Fraternal twins are the result of the fertilization by two sperms of two different ova which arrived in the Fallopian tubes at approximately the same time. Usually only one ovum appears, thus giving rise to only one child. Since the twins are the result of separate eggs and sperms, it is not to be expected that they should be too much alike.

In identical twins, a single fertilized egg divides in two to produce two individuals, each possessing the exact same embryonic material derived from the single egg. Thus, the twins are exact genetic copies of each other, even to the extent of being of the same sex. Although they each have their own separate umbilical cord, in most cases they share one placenta. Fraternal twins also have their own umbilical cords, but each has a separate placenta.

It is interesting to note that twins have a better chance of occurrence among prior mothers of twins and seem to run in females of the same family, being passed on from mother to daughter. Usually twins are born prematurely by about three weeks.

WHY DO WE TURN WHITE WITH FRIGHT?

Just as there are nerves that dilate blood vessels, there are nerves that contract them. These latter nerves are called vasoconstrictor nerves. During a sudden scare or fright, the face takes on a pallor. The constrictor nerves become stimulated and cause the blood vessels to contract. Thus, the supply of blood is lessened, and the result is a pallor.

DID THE REFRIGERATOR HANDLE GIVE YOU A SHOCK?

When electric charges collect on a body and remain there, they are called static or stored charges, and give off static electricity. This is what makes static electricity different from continuously flowing electricity. When a person walks on a rug for a period of time, the friction causes him to acquire static electricity in the form of negatively charged electrons on the body. These electrical charges on the body, especially on the extremities such as fingers and toes, are anxious to be neutralized with a positive charge. When a person touches the aluminum handle of the refrigerator to open it, the handle, being both uninsulated and a good conductor of electricity, causes a spark when the electrons leave the body and enter the handle. The body thus becomes neutral again.

WHY IS SALT WATER EASIER TO SWIM IN THAN FRESH?

Salt water is more dense than fresh water, meaning it weighs more per cubic foot than fresh water. This gives your body more buoyancy so you can float a little higher than in fresh water. It is this extra upward pressure of buoyancy in salt water that makes swimming easier.

WHY IS THE HYPOTHALAMUS SO IMPORTANT?

In the brain, the lower part of the diencephalon (above the inner regions of the nose) is called the hypothalamus. A hanging stalk from this area supports the pituitary gland, thus connecting the hypothalamus to the pituitary gland both in function and in structure.

Scientists believe that the functions of the hypothalamus are as follows:

1. Regulates the heartbeat.
2. Regulates the rate of breathing.
3. Regulates body temperature.
4. Regulates cycle of wakefulness and sleep.
5. Is believed to act as a biological timer to start the process of sexual maturity.

WHAT HAPPENS WHEN BODY TEMPERATURE RISES AND FALLS?

The chemical reactions which build up in our bodies occur best at the temperature of 98.6°F. (37°C.). But a higher temperature could break up delicate and complicated compounds called enzymes. A lower temperature would result in a lessened rate of chemical activity which could lead to serious complications. A little difference between your awakening temperature and your temperature at 6 P.M., which can vary as much as two degrees, means nothing. But convulsions may occur (though not always) at about 106°F. and brain damage occurs at 108°F. At the other extreme, a person loses consciousness at 91°F. and heart problems develop at 83°F.

WHY AVOID TREES DURING A LIGHTNING STORM?

A tree is a conductor of electricity, and instead of being a protector, it is a great source of danger. The sap of the tree is a liquid conductor of electricity and therefore makes it dangerous to stand under for protection. It is as dangerous as being in a swimming pool during a lightning storm.

WHAT IS STRIP MINING?

In the state of West Virginia there exists an abundance of coal in the mountains instead of the usual underground source. The coal lies in layers between topsoil and shale, and to get at it, machines scrape off the earth and rock, and the coal is removed. The debris is pushed over the side of the mountains as the machines go along. Strippers then bore deeper into the mountain to extract more coal. The net result in regard to appearance is a series of giant terraces marking where a layer of coal was scraped up. No trees or other vegetation can be seen between terraces. Since the state's laws do not seem to have much power to prevent this, help is being sought by the environmentalists and conservationists to get the federal government to intervene by means of laws enacted by Congress.

WHAT IS RAIN?

When the heat of the sun warms the air and makes it rise, it carries with it some moisture. As it rises, the moisture-laden air expands and gets colder. Soon it is visible as a cloud, and inside it, drops of water begin to form. As the drops are swept upward by the rising air, they become heavier and fall as the first raindrops. The cloud gives out its moisture content as rain, and is usually welcome by everyone as an essential of life.

HOW ARE CORAL REEFS FORMED?

A coral reef colony starts when a tiny larva anchors itself on a firm surface. From the upward end a set of tentacles soon grows out around its mouth. It is now a polyp or full-fledged coral animal. It is saclike in form now, and for protection takes the dissolved calcium from the water and deposits it beneath and around its body to form a cup of calcium carbonate known as limestone. Its tentacles extend from the opening at the top of the limestone cup and function only in the dark.

As the polyp grows, it soon divides or buds, forming two identical polyps. Soon each is enclosed in its own stony cup. They grow most rapidly where water is clear, salty and warm, and soon appear like branches of a tree. They remain associated with each other, thus forming colonies. The combined limestone-forming activity of these colonies produces masses of large size in various beautiful forms and shapes.

Another group, the soft corals, grow in a similar way, only the soft corals are more flexible in relation to their skeleton. They are typified by the graceful movements of the "sea fan," which gently wafts back and forth with the undulations of the waves above it. The hard coral polyps have independent digestive systems, and each polyp must fend for itself. The soft coral's polyps are interconnected so that food obtained by the tentacles of one is digested and distributed to all the others in the colony.

The living polyps cover the outside of these masses, and give rise to the beautiful colors of the reefs. These corals make up the coral reefs and help build up the foundations for hundreds of tropical coral islands, atolls and barrier reefs. The source of color of corals is the algae which live inside the polyps' tentacles. These varicolored algae give the coral colony its beautiful hues.

WHAT CAUSES SNOW?

Snow is formed when water vapor in clouds is turned into moisture at a temperature below freezing, which is 32°F. It does not always reach earth in its original form, which is a transparent crystal too small to be seen. Sometimes the ice crystals are partly melted and reach the ground as sleet, but when entirely melted, they fall as rain. When snow reaches the ground, it is a composite of many crystals joined to create an exquisite hexagonal design, no two of which are alike. The air currents they encounter, in their fall to earth, helped to join them together in these beautiful designs. It is estimated that about one inch of rainfall is the equivalent of ten inches of snow. But it has never snowed on more than one-third of the earth's surface, and parts of southern United States have never seen the beautiful sight of crystal-white snow drifting to earth.

WHAT IS A TANTRUM?

One day, a little boy in a department store refused to dismount a hobby horse, even for the store police. Finally the manager whispered in his ear, and the child got off at once. When the mother asked the manager what he said, the manager replied, "I told him to get off or I'd knock him off." Another child did the same thing, and when the manager whispered in his ear, the child bit him, and then wrestled his mother to the floor as she tore him from the horse. Then he lay there kicking and screaming, but no one could do a thing about it. The second child was different because he was prone to tantrums—fits of bad temper. As his rage increased, it got out of control of his central nervous system, and no amount of reasoning could help. He was not functioning with his brain.

Tantrums are frequent in a particular child. He may want a good audience, try to get his parents under his thumb, or he may be emotionally upset. Neither pampering, giving in, nor physical punishment can help. Parents must sit out the raging storm and wait for the child to succumb to fatigue and sleep. As the child matures, tantrums usually subside.

HOW IMPORTANT IS THE THYMUS?

Not very long ago the thymus gland was considered a nonproductive evolutionary appendage with no known function, much like the appendix. Today it is given a role of great importance.

From an original weight of two-thirds of an ounce it remains in our body till it is about four ounces at puberty. From then on it begins to shrink until it is about one-third of an ounce at age 50. This yellow-gray blob of tissue, about the size of a man's wrist watch, lies atop the breastbone. From infancy to puberty it is the chief defense mechanism against body infection, by producing lymphocytes and supplying a hormonal stimulus to prod into activity the spleen, lymphatic system, and other organs.

If there's a possibility of infection, no matter how small, the lymphocytes produced by the thymus pour out the antibodies to attack and kill the invaders. These invaders may be bacteria, viruses, fungi or foreign tissue. There are millions of different antibodies, a special one for each disease, which come forth to fight off disease. At about age 50, the immune response slows, the thymus no longer produces lymphocytes, and the defense mechanism begins to break down. This is attributed to the shriveling up and almost complete disappearance of the thymus. This may be a part of the aging process, and it is hoped that the recently discovered thymus hormone called thymosin may some day be used to halt the aging process and keep us young a little longer.

WHAT IS THE STORY OF THE ELECTRIC LIGHT BULB?

When the electric light bulb was first invented by Thomas Edison in 1879, he used a carbon filament in the globe. Although it glowed beautifully when the current was applied, it soon evaporated or, as we say, burned out, because of the extreme heat of oxidation. In order to reduce the speed of oxidation on the carbon filament, he got rid of the oxygen by sucking the air out and creating a vacuum in the bulb. Now the filament lasted a much longer time, and for several years carbon was used as the filament.

However, Edison knew that he could get a longer-lasting brilliant light if he had a better filament. So he started a worldwide search until he came upon the rare silver-white metallic element known as tungsten. This was exactly what he sought because it had a very high melting point (6,143°F.) and a great resistance to oxidation. It is also one of the hardest known metals.

To this day tungsten is used as a filament in electric bulbs, but inert gases which do not combine with other substances are substituted for the vacuum. This further reduces the rate of evaporation of the tungsten and gives it a longer life in the tube.

HOW DOES A STARFISH EAT A CLAM?

Did you ever watch a shucker at a fish place open clams and oysters? Seems easy, doesn't it? Well, if you ever tried it, you'd know it takes months of practice to learn this business. At home, if you ever tried to open a dozen or so, I'd wager you could not do it even with a hammer, chisel and carpenter's vise to hold the clam. I mean, without breaking the shell and squashing the clam. Well, by now you wonder how the starfish does it, because he doesn't use a single tool and a clam or oyster is his favorite dish. For this reason starfish are the curse of oyster breeders.

On the underside of the starfish are grooves running from the central mouth out to each arm. From tiny holes in these grooves come forth little suction tubes whenever a clam or oyster is caught. The starfish uses its tube feet with these sucking discs at the end, in order to open the two halves of the shell. By exerting a great suction force in opposite directions, the shells of the clam are finally opened. For a long while the clam resists with its inner strong muscle, but the starfish, being more powerful and patient, must win. Eventually the clam relaxes, gives up and surrenders its life. The starfish then pushes its stomach through its mouth opening and into the open clam. Then, with the aid of its own special clam juices, it begins to digest the delicacy.

WHAT FORCE MAKES THE EARTH ROTATE?

The fellow who wrote the song "Stop the World, I Want to Get Off" didn't realize what an impossible dream that was. The world keeps rotating without a stop at 1,000 miles per hour (at the equator) and revolving around the sun at a speed of 66,600 miles per hour.

We know that the earth is going through these motions, and we can prove it. But what the force is that makes it rotate, no one really knows. There is no answer that can be backed by fact or definite proof, but there are several theories, one of which says that the sun is a mass of spinning gas from which the earth broke off. When the earth broke off and spun itself away from the sun at tremendous speed, it kept on spinning on its imaginary axis and is still doing so to this very day.

HOW IS DEW FORMED?

The heat of the sun causes moisture from the earth's surface, from bodies of water and from plants to evaporate into the warm air. After sunset, the air and the earth grow cold. Then the moisture in the air condenses and forms drops of dew on grass and other plants.

WHAT ARE FRECKLES?

Everyone, except albinos, produces a certain amount of melanin, which is a dark pigment that absorbs ultraviolet rays from the sun. This pigment is produced by certain cells located in the top layer of the skin, called melanocytes. These yellowish-brown spots on the skin usually are on the parts of the body that are exposed to the sun, such as the arms, face and neck.

The pigment, which consists of tiny granules in the cell, gives the skin its color. Freckles are localized patches of these pigment cells. These groups of active melanocytes, surrounded by groups of less active melanocytes, produce the islands of pigment known as freckles. The condition is considered to be hereditary.

DO WE HAVE TO SLEEP?

If you know what's good for you, you had better get your full quota of sleep. Every person has got to have periods of sleep in order to regain the lost energy in his muscles. Some need more, some need less. This depends upon the nature of the person and the nature of his work. Athletes may require more sleep than elevator operators. Even that depends upon the days and hours they play or work. Even their ages make a difference.

Psychiatrists have made studies and performed experiments to show that a person kept awake long enough will start having all the hallucinations of a person with a serious mental illness. After five or six days and nights without sleep, some volunteers in the experiment "saw smoke coming through the air conditioner" or saw people who were not actually there, or in general reacted the way seriously schizophrenic patients behave. Except for the difference in time for the reaction to take place, everyone showed the same schizophrenic results. Those who did not reach this stage in the experiment were the ones who fell asleep before the experiment was over.

To a great degree, we respond to our work, our neighbors and friends according to the well-being of our body and mind. All life responds to the natural rhythm of the seasons and the changes of day and night. Even plants and flowers have to have their periods of sleep and rest.

WHAT GIVES FLUORESCENT LIGHTS THEIR COLOR?

A fluorescent light is a long glass tube with an inside coating on the glass. The coating may be one of many substances depending on the color of lighting desired. It might be phosphor, zinc silicate, calcium tungstate, zinc beryllium silicate or magnesium tungstate. The tube is then filled with mercury vapor and sealed. The conduction of electricity through the mercury vapor in itself does not give light, but gives off ultraviolet rays which are used to cause fluorescence or glowing in other substances.

WHAT REGULATES BODY TEMPERATURE?

The two temperature senses, cold and warmth, correspond to distinct nerve endings. Both occur in the skin, but their distribution is not by any means identical. Areas containing many warmth-sensitive endings often contain few cold-sensitive endings and vice versa. A strange result is the touching of hot water or dry ice. Both give a sensation of burning.

When the body is too hot, the hypothalamus signals the blood to rush to the skin to cool off in the capillaries. At the same time a signal goes to the sweat glands to send perspiration through the pores. Thus the body gets cooled when the perspiration is evaporated by the breeze. On the other hand, if the hypothalamus finds the body too cold, it signals the body to slow down the blood flow to the skin, retain most of the blood in the inner organs, and also signals the sweat glands to refrain from perspiring. A marvelous engineer lives in our brain.

WHAT IS HAIR AND WHY IS IT STRAIGHT OR CURLY?

Except for an occasional discovery in a plate of soup, hair is usually found on all mammals. In a slightly different form from ours, it is called fur on animals, but like human hair, it is also protein. Hair has an overlapping layer of fine scales, arranged like the shingles on a roof, except that the free ends point upward. The part outside the skin is called a shaft, and the portion below is called the root. The root is pear-shaped and extends down into a pocketlike sac called a follicle. A slight projection of the follicle is called the papilla, where the blood vessels lie that carry life-giving material to the hair.

Now, you may ask, why do some people have straight hair, some curly, and some kinky? It's a good question, and if you'll forgive a pun, we won't split hairs over the answer. The type of hair you have is purely a hereditary factor. It was all predetermined before you were born by the genes in the chromosomes.

The shaft of hair in most races is usually round, but

in some persons and races it is flattened. Flat hair has a tendency to curl because it grows unequally at the different angles of the overlapping scales. The kinky hair of the black races is as flat as a ribbon. But flat or round, sparse or bushy, red or blonde, our modern-day tonsorial artists can defy the wishes of the strongest chromosomes.

CAN THE SUPERSONIC TRANSPORT BE DANGEROUS TO US?

The earth has a protective shield of ozone which keeps out the dangerous ultraviolet rays of the sun. At supersonic speeds, some scientists believe that the ozone shield could be sufficiently shattered to permit the ultraviolet rays to penetrate the earth's surface. They estimate that this could cause skin cancer in about 8,000 people each year and kill about 300. This is a warning that came from a 46-page scientific study made by the Environmental Studies Board, a special panel of the National Academy of Science. Today, supersonic transports are in operation, crossing the Atlantic several times a week. Only the future will tell whether the warning from the environmental board should have been heeded.

HOW DOES A MOUNTAIN BECOME A GRAIN OF SAND?

All rocks are in a constant state of gradual erosion. This applies to mountains as well. When water gets into mountain crevices and freezes, the expansion process breaks off large chunks of rock. These in turn break into small pieces as they tumble down the mountain-side. After millions of years of exposure to the whims of rain, storm, wind and seas, they will gradually take the form of tiny grains of sand.

WHY ARE HEATERS PLACED LOW AND COOLERS HIGH?

Because heat makes molecules move and bounce faster, there are fewer molecules in warm air than there are in cold air. This makes warm air much lighter than cold air and therefore it rises, while cold air, being much heavier, falls downward. In order to get the most beneficial use of a heater, it is set on the floor to permit the warm air to rise and heat a room. The reverse is true of a cooler or an air conditioner. It is set at a higher level to permit the cool air to fall downward and cool the room. It is only because the tiny molecule of air is so dependable and agile that we can put it to use to warm us up or cool us off.

WHAT CAUSES A TORNADO?

Although we don't know exactly how tornadoes form, this is the prevailing theory. When a mass of cold air moves easterly at an increasing rate of speed across the southeastern part of the United States, the cold mass creates a sort of shock wave that moves ahead of the mass. This shock wave causes air pressure to rise rapidly. If a mass of warm, damp air moves northward from the Gulf of Mexico, a second shock wave will be produced and continue northward. When the eastward moving wave and the northward moving wave meet, both are diverted from their paths and a whirl of air results.

A tornado is one of the most violent of all weather phenomena. This whirling mass of air is so strong its winds cannot be measured with modern equipment, but have been estimated at 500 miles per hour. Its suction force has been known to lift automobiles into the air and lift a complete roof from a home.

They are difficult to predict, and usually occur in spring and summer when the lower layer of air is warm and moist and the upper layer is drier. Every state reports at least one tornado a year, but eastern Kansas seems to be the center of the tornado region. For 35 years over 15 tornadoes for each 50 square miles of Kansas, Nebraska, Iowa, Missouri and Oklahoma have been recorded. Florida also has a high incidence of tornadoes. When they occur at sea, they are called waterspouts.

WHY DO WE HICCUP?

I doubt if anyone really knows why we hiccup, or what good it really does for you. Other than having just nuisance value in simple cases, and in extreme cases sending some people to the hospital for a period of weeks, a good reason for its occurrence is hard to understand. It is much easier to explain a yawn, a blush, or a sneeze, but a hiccup is a real toughy.

But if you wish to know from the mechanical standpoint why you hic and hic, try to follow me. The muscle that contracts and expands your chest cavity is the diaphragm, and certain nerves control its contraction and relaxation. Sometimes, for some unknown reason, a sudden nerve mixup causes short spasmodic up-and-down movements instead of the natural slow ones. These quick spasms cause your vocal cords to snap shut, and your windpipe to close. This brings on the hiccuping.

By now you can see there's really no good purpose to the hic, which can come most unexpectedly at any time. You could be eating food, drinking alcohol, or suddenly swallowing air, when along comes that hic, hic, hic. Everyone has a home remedy for the hiccups, but it seems that only nature, who brought it on, can help it go away.

WHAT ARE PULSARS?

There are certain stars in the sky that are so infinitely far from our earth that our ordinary telescopes cannot reach them. But we have discovered that they are positively there. How? Well, some of these tiny stars that we have named pulsars or neutron stars, about 2,000 trillion miles away, send out regular impulses or signals which are recorded every 1¹/₃ seconds. In spite of their unbelievable weight (a billion billion billion tons) and a diameter of twelve miles, they make one revolution every second. A scientist by the name of Burnell discovered these about ten years ago, and up to this date about 150 pulsars have been tracked. To search out these stars, a new method was created, using radio waves to detect impulses from faraway objects. This radio-wave telescope is about four and a half acres in size (about four city blocks) and the recording device for the signals is much like the cardiograph machine used in hospitals.

Naturally, very little is known about them at this time, since we have not yet perfected a telescope capable of reaching them. This is another one of science's interesting problems which we hope some day to solve.

HOW FAR IS HOME FOR THE HOMING PIGEON?

Although science cannot explain the phenomenon of the homing pigeon, it doesn't mean that the subject should not be discussed. We know that the bird is swift, courageous and intelligent, but how it finds its way home is beyond us. Some scientists try to link the answer to the sun, the North Star, magnetic fields, etc., but all of these are theories. How can we explain a flight of 7,200 miles from Arras, France, to a city in China? It's like going from New York to California and back. Short distances of 200 or 300 miles are often performed in competition among homing pigeon fanciers. The mystery of these flights, both short and distant, is that the covered territory is so vast and strange that the return home cannot be attributed to the power of visual memory.

Starting with short training flights when the pigeons are young, the distances are slowly increased. As early as 3,000 years ago, the Egyptians and the Persians used them to carry messages from city to city. The messages were rolled into capsules and attached to their legs. They were used by the Romans in their numerous wars and by the French in the Franco-Prussian War. But the Germans retaliated with hungry hawks to catch them. During World War I, an intrepid and wounded pigeon completed its flight with vital information that saved the Lost Battalion. Its name was "Cher-Ami," and to the memory of its bravery, I dedicate this page.

WHAT IS A LASER BEAM?

In every substance the atoms are arranged differently as they fly around their different orbits. This is what makes all materials unique. For example, in the ruby, the atoms are arranged in such precise geometric fashion that when they are stimulated by a current to expand and contract their orbits, the resulting ray of light is a laser beam. One end of the ruby is first highly polished, smoothed and given a mirror plating.

In ordinary incoherent light, the rays come out of the source in a straight line, in helter-skelter fashion, in all directions. But not so with this laser beam. Its light is produced by the release of energy from the orbiting electrons which alternately collapse and expand their orbits.

Like an array of military soldiers, each orbit expands and collapses in precise uniformity, thus creating a pumping action which causes the released energy to emerge in a phase, or in step. This is called coherent light, different from ordinary incoherent light because the rays come out in the same polarized plane in the same frequency (color) as opposed to the helter-skelter fashion of light rays.

WHAT IS MULTIPLE SCLEROSIS?

If you were watering your lawn and the hose you were using had many breaks and leaks in it, you'd hardly get enough water to the grass. Well, the nerves of your central nervous system are composed of fibers that are covered by a myelin sheath to see that impulses are not lost on the way to the brain. They are like electric wires, insulated by a covering to see that the current reaches its destination without any loss en route. Of course, you know what would happen if there were cuts and breaks in the wire. Apart from possible serious damage, the current wouldn't reach its intended destination. A nerve impulse is like an electric current whose destination is the brain center, and whose purpose is to inform it of some important happening. It could be an idea, a response, or even a sense of pain or joy.

If the myelin sheath covering the nerve fibers disintegrates into a patchwork of acute lesions (like the damaged water hose), the impulses get short-circuited between the brain and the intended body recipient like the arms or legs. Thus, the control of the nervous system is sadly disrupted and crippled. This is multiple sclerosis.

Strangely, no other animal suffers from this disease; it is strictly a human affliction. Strange too is the fact that although this disease has been studied in depth since it was first described 150 years ago, the cause of multiple sclerosis has never been established. It is still a mystery today.

WHAT INSECT IS THE WORLD'S GREATEST ATHLETE?

Can you imagine a human capable of a 600-foot broad jump and a 360-foot (36-story) high jump? Well, that's what it would amount to if humans had the proportionate capability of a tiny flea, less than an eighth of an inch in size. In nine leaps you could do a mile. Who would need cars, buses or trains? In 29,000 leaps you could be in California. Who would need a plane, unless you had a sprained ankle? Judging from their prowess as broad jumpers and high jumpers, the fleas are the greatest athletes in the world. I mean those on your pet cat or dog. Did you ever try to catch one? They can do a thirteen-inch broad jump, and an eight-inch high jump.

This shouldn't surprise us one bit, because scientists tell us that fleas have been on this planet for over 40 million years. With all those years of practice, jumping high and broad, I can see why they are so expert at evading flea collars and soapy baths. These rascals thrive on the warm blood of your dog or cat, and even yours if they can sneak into your underwear and get away with it. Then, after a bellyful, they go out to propagate their species.

Considering its tiny size, a life span of 2½ years is quite long. Too long, if you ask your dog. Fleas are also selective about climate. They don't like it too cold or hot, wet or dry, but just right, a room temperature of about 65 degrees F. That's why you'll see them high jumping in your rug. The little eggs that roll off your pet's fur land there and go through a larva stage, and later emerge from their cocoons as adults, to bite and jump and plague us for another 40 million years. There's also a flea known as a driller who literally gets under your skin. But we'll catch him another time.

HOW DO SMOKE DETECTORS WORK?

These electronic smoke detectors are remarkable because one can hardly believe that a tiny molecule of smoke, something you cannot even feel, is capable of setting off a loud alarm to warn you of possible danger. How does this happen? The same principle described in the working of the electronic eye door on another page in this book applies to this gadget. These alarms can sense a fire long before the flames appear, and send out an ear-piercing signal. Once installed, they can be interconnected to different rooms anywhere in the house. The gadgets work by electricity or by battery, using a photoelectric tube or cell. The tube can "see" smoke because a tiny light beam spots the smoke particles that come through the holes in the detector. When the smoke that enters the detector breaks the light beam inside, the reflected light rays trigger the cell which in turn signals the siren to go off.

There are several variations of this device. One is a

gas detector, and one has an "ionization" chamber that smells fire and ionizes air in the small chamber, thus creating an electric current sufficient to set off the alarm. These devices have proven to be so effective that many states are beginning to require them by law.

DO DOCTORS EVER GIVE A PHONY PILL?

You bet they do, and have been doing so for as long as prescriptions have been written. It has been well known in the medical profession that when a patient pays a doctor for a visit, unless he or she receives a prescription for some kind of medication, he feels that the doctor is lax or doesn't take his complaint too seriously. So, to those many patients who he feels need no medication other than his advice, he prescribes this phony pill, called a placebo in the profession. The name is derived from the Latin meaning to please, and the pill is nothing more than a capsule of powdered milk and sugar.

There is substantial evidence that placebos have really had therapeutic results, that they can be as effective as, or even more effective than, the active drugs they replace. No drug is without some side effect, and a good doctor does not wish to compound the symptoms with additional problems. Just as patients can be influenced to undergo surgery, so they can be influenced to take a series of placebos. Of course, the pill itself is not the cure; it is the biochemical changes aroused by the mind-set to overcome the illness that is really the cure.

Since illness is an interaction of mind and body, the placebo can set the mind to work to see that the body gets well. Of course, placebos don't always work, nor are they prescribed where the doctor feels the patient needs the real capsule to get results. But in many cases he does prescribe placebos because, to put it simply, it is a matter of mind over matter.

ARE YOU WORRYING ABOUT CHOLESTEROL?

At last it is nice to know that regardless of whether the cholesterol count in your blood is high or low, you need not worry about it any more. In the first place, doctors are concluding that it's due to heredity, and secondly, that it's not much of a problem in bringing your cholesterol around to normal. A proper diet—that is, staying away from meat, shellfish, dairy products and eggs—and just doing a little exercise each day, and there you have it, all in a nutshell and no fee for the advice.

Cholesterol is a fatty substance made by the liver and derived from the aforementioned foods. But how about people who eat all this stuff, have no high cholesterol reading, and who live long lives free from heart disease? Studies under an electron microscope showed that cholesterol is carried through the blood by molecules called lipoproteins. Now here is the payoff. There are good ones and bad ones, and this conclusion came after 28 years of study of 5,127 people by Dr. William Castelli, director of laboratories for the Framington (Mass.) Heart Study.

The low-density lipoproteins (LDLs) are the bad ones. They pick up the excess cholesterol not used in your daily metabolism, and deposit the fatty molecules on the lining of coronary arteries. But the good guys, the high-density lipoproteins (HDLs), pick up the excess cholesterol from the bloodstream and bring it back to the liver to be excreted from the body. How come goodies and baddies? We don't know. But at least, by a simple test, we can tell whether you must diet and exercise, or whether you are one of the lucky ones who could eat like there was no tomorrow without a worry about cholesterol.

WHY IS THE OCEAN BLUE?

The seas are blue for the same reason that the sky is blue. The blue color is caused by the scattering of sunlight by tiny particles in the water. Blue light has a short wavelength, as compared to the wavelengths of other colors, and is therefore scattered more effectively than the others. However, the microscopic floating plants may give the water several different shades, such as green (in tropical waters near the coast) or a brownish color near shore silt or suspended sediment.

WHAT IS INDIA INK?

This ineradicable, waterproof and absolutely black pigment was used by the Chinese around 2600 B.C., and is still used and made the same way today. It is made by mixing lampblack with glue or gum. The paste is then dried and later mixed with water. Artists all over the world use this ink for drawing lasting sketches that never fade.

HOW DID A THREE-INCH FISH GET INTO THE SUPREME COURT?

Everybody knows, or should know, that you cannot take a case to the U.S. Supreme Court unless a constitutional question is involved—such as an invasion of a person's civil rights. So if you owned a house right smack in the center of where the government planned to build a 119-million-dollar dam, you could be 100% sure that your house would be knocked down, even if you had the best lawyers in town and it took a one-ton wrecking ball to do it. Why? Because the right of eminent domain gives the government priority over your rights to any property it needs to use in any government project. Of course, you will be paid for the property in accordance with an appraiser's evaluation, but you would have no chance of getting your case in front of the "nine old men" in Washington.

But you'd better believe that a little three-inch fish called the snail darter made it to the hallowed confines of the courtroom and won a mighty battle. The reason: In 1973 Congress passed a law forbidding any act that might harm an endangered species. Only about 1,600 snail darters exist in the world, and it was believed that building a projected dam would flood their habitat and wipe out the species. Could there be a greater invasion of civil rights? So the nine men declared that a three-inch fish could prevent the erection of a seventy-foot dam. Who can deny the wisdom of a country where a little fish can come into the highest court?

HOW DOES A FUSE PREVENT A FIRE?

A fuse is an important safety device used in all electrical circuits to prevent a circuit from becoming overheated. Just as a hot water boiler needs a safety valve to prevent an explosion when too much pressure is accumulated, so must an electric circuit have a fuse in its system to prevent a fire.

A fuse contains a small metal strip that melts at a lower temperature than the other wires in the circuit. The fuse receives the same amount of current as the house circuit, and if one places too much of a load on the circuit by using too many electrical appliances at the same time, the heat causes the metal in the fuse to melt and break the circuit, thus preventing a possible fire due to overheated wires.

Of recent innovation is the device called a home circuit breaker, which serves the same purpose as the fuse. When a circuit is overloaded, a switch flies up and breaks the circuit. When the excess load is taken off the circuit, the switch can be pulled down to start the flow of current again. The home circuit breaker is presently replacing the fuse in new homes.

HAVE CHROMOSOMES AND GENES BEEN SEEN BY MAN?

In 1956, when proper techniques were developed to make accurate observations of human chromosomes, it was possible not only to count, but also to identify, each of the 23 pairs of chromosomes contained in the human cells.

As for genes, researchers studying the midge, a small fly, photographed the chromosomes taken from the salivary gland cell. Although they did not know exactly which genes related to the molting process, they treated the chromosomes with a hormone that induces molting. Soon, certain genes were seen to puff and rise as if reacting to this hormone. This shows to some minor degree that the genes of the chromosomes react to certain hormones.

WHY IS ADRENALINE IMPORTANT?

Adrenaline is a hormone produced by the two ductless adrenal glands which sit atop the kidneys in mammals.

During exercise, and especially when there is fear, anger, danger or sudden muscular activity, the glands are stimulated and larger amounts are discharged into the bloodstream. This stimulates the sympathetic nervous system with the following results:

1. Blood pressure is increased quickly and markedly.
2. Respirations are quicker and deeper.
3. Smooth muscle tissue is contracted.

Adrenaline is the most rapidly acting circulatory stimulant of great power, but because its action is brief, it is used only in emergencies. It is used in cases of failure of the normal heart as may occur following electric shock or during a surgical operation. Then it is given in dilute solution directly into the bloodstream. Many lives have been saved by this emergency treatment.

IS THERE ANYTHING BENEFICIAL ABOUT A VOLCANO?

It's hard to believe that rocks can be melted or turned into a gas or vapor. Yet that is what happens when the heat of melted magma, twenty to forty miles below the earth's surface, seeks an outlet to the surface. When the accumulated pressure can no longer be contained, the hot magma pushes its way up the passageway or conduit, melting all the layers of rock as it comes up. The molten rock and gases, after reaching the surface, become lava, which flows over the outside of the crater until it hardens. The eruptions are caused by the action of accumulated gases in the upper part of the conduit.

On August 24 in A.D. 79, a famous eruption took place on the slopes of Mt. Vesuvius, near Naples, Italy. Three cities were destroyed and buried under lava deposits, and remained so for nearly 1,700 years. In 1748 a buried wall was uncovered which led to the eventual discovery of the ruins of Herculaneum, Pompeii and Stabiae.

Yet volcanic eruptions can be beneficial to mankind. In Italy, Sicily, Iceland, Chile and Bolivia, volcanic steam is used to run plants that create heat and power. Pumice, a by-product of lava, is used as a grinder and polisher and in the building of roads. Sulfur, a volcanic product, is used in the chemical industry, and lava is used as building material. Soil from decayed volcanic material is rich in minerals and produces good harvests in the agricultural fields of the Hawaiian Islands. So you may conclude from all this that nature, in a way, tries to compensate for some of the damage it does.

WHAT IS THE DIFFERENCE BETWEEN DC AND AC CURRENT?

DC means direct current and it stands for the electricity that comes directly from the source of origin, such as that from a storage or flashlight battery, and goes to the point of use without any fluctuations or alternations.

AC stands for alternating current. This means it does exactly what the name implies. It alternates its direction and goes back and forth at the rate of 60 times per second, from its point of origin, the power company's electric generators or dynamos, to its point of use, the home or factory.

The alternating speed is so fast that it is not noticed at its place of use. The reason that the current produced by these generators is alternating is that moving coils of wire cut the lines of force of a magnet. The more lines of force cut by the coil, the stronger is the current.

In the electroplating industry, where direct current is used, the electricity is produced by a generator whose collecting rings are so arranged that they only give off a direct current.